Mobile Farmers

An Ethnoarchaeological Approach to Settlement Organization among the Rarámuri of Northwestern Mexico

by

Martha Graham

INTERNATIONAL MONOGRAPHS IN PREHISTORY

Ethnoarchaeological Series 3

Printed in the United States of America

ISBN 1-879621-16-9 (Paperback)
ISBN 1-879621-17-7 (Hard Cover)

Library of Congress Cataloging-in-Publication Data

Graham, Martha, 1957–
Mobile farmers : an ethnoarchaeological approach to settlement organization among the Rarámuri of northwestern Mexico / by Martha Graham.
p. cm. — (Ethnoarchaeological series ; 3)
Includes bibliographical references.
ISBN 1-879621-17-7 (hard cover : acid-free paper). — ISBN 1-879621-16-9 (pbk. : acid-free paper)
1. Tarahumara Indians—Agriculture. 2. Tarahumara Indians—Land tenure. 3. Guarijío Indians—Agriculture. 4. Guarijío Indians—Land tenure. 5. Land settlement patterns—Mexico—Rejogochi. 6. Subsistence economy—Mexico—Rejogochi. 7. Ethnoarchaeology—Mexico—Rejogochi. 8. Dwellings—Mexico—Rejogochi. 9. Rejogochi (Mexico)—Social conditions. 10. Rejogochi (Mexico)—Antiquities. I. Title. II. Series.
F1221.T25G73 1994
338.1'089974—dc20 94-6297
CIP

This book is printed on acid-free paper. ∞

International Monographs in Prehistory
P.O. Box 1266
Ann Arbor, Michigan 48106-1266
U.S.A.

Tel. (313) 761-9068

Table of Contents

Preface

One important aspect to understanding a cultural system archaeologically is the accurate identification and explanation of settlement organization. Our interpretations of land use practices have implications for reconstructions of population, levels of technological innovation, basic subsistence, social and religious organization. Traditional methods of settlement pattern studies were designed to *describe* prehistoric cultures rather than to address issues of the *organization* of cultural systems. Thus, these methods not only prevent us from identifying important variability in the archaeological record, but also from interpreting that variability in meaningful terms of cultural process.

My ethnoarchaeological work among the Rarámuri (Tarahumara) of northern Mexico investigates one aspect of such variability, posing the relatively simple but important question: How can archaeologists identify mobile agriculturalists? In order to address this question, I consider the archaeological implications of residential mobility that can be drawn from an investigation of Rarámuri settlement organization. I compare the Rarámuri, who are mobile agriculturalists, with the Guarijío and the Tuxtleco, non-mobile groups, to highlight the ways people with different settlement organizations use their residential space. By looking at aspects of residential sites that are observable in the material record and that have meaning in terms of cultural adaptation, I identify several approaches that help answer the question posed above.

Mobile agriculturalists use space in a more generalized way than non-mobile groups. The latter use more specialized activity areas for carrying out many activities at their residences. Nonmobile agriculturalists also invest more in facilities than do non-mobile agriculturalists, such that some kinds of facilities are fixed permanently in one place. Among more mobile agriculturalists, these facilities are part of the site furniture left at the residence during periods of abandonment, but can be moved from place to place at the residence. Non-mobile and mobile agriculturalists also emphasize different maintenance behaviors at residences. Non-mobile agriculturalists use strategies resulting in areas of more concentrated refuse; the trash zones of mobile groups tend to be diffuse. These differences in spatial arrangements, permanence of facilities, and refuse areas have material consequences that we might identify in the archaeological record. Our ability to monitor aspects of mobility in the archaeological record will help us develop a better understanding of settlement organization.

Funding and support for the research on which this book is based come from a variety of sources. The fourteen months of fieldwork were supported by grants from Sigma Xi, the Tinker Foundation of the University of New Mexico Latin American Institute, and a Dissertation Improvement Grant from NSF (BSN-8611525). A Graduate Student Fellowship at the Smithsonian Institution in 1986 provided me the opportunity of 12 weeks' introduction to the Rarámuri language under Bill Merrill's instruction, and a Smithsonian Pre-Doctoral Fellowship provided me with access to additional Rarámuri data from Rejogochi.

I am indebted to the Rarámuri of Rejogochi for taking me into their community and families and for their good-natured acceptance of my frequent questions, surprising ignorance, and strange behaviors. A special thanks goes to the children of Rejogochi; they often understood what their parents could not— that I was truly ignorant of the basics of Rarámuri life. As their parents taught them, they taught me, and for that they were some of my best teachers as well as good playmates. The *chabóchis* (non-Rarámuri) around Basíhuare and Rejogochi were always friendly and supportive. I am especially grateful to Father Luis Verplancken, Margarita, Francisco, and Jorde.

In the United States, Karen Dohm, William Latady, Alexandra Roberts, and Lawrence Todd exchanged ideas and information relating to this book. I especially appreciate Karen's comments on this manuscript. The research for this book was conducted while I was a graduate student at the University of New Mexico, and the results written originally as my dissertation. Special thanks go to Lewis Binford for his encouragement and his example in doing site structural research.

I owe a special debt to William Merrill for his support and encouragement. Bill freely allowed me to use his data and other resources, and Figures 5 and 8 come from his own 1977-1979 field data. I also appreciate his comments on this manuscript. The information on Household 8, especially Figures 21 and 32, come from the 1983 Tarahumara Ethnoarchaeology Project, when I served as Robert Hard's assistant on the Tarahumara Ethnoarchaeology Project. I am grateful to Bob for letting me include them here, and for first introducing me to the Rarámuri.

This research could not have been accomplished without the use of additional data acquired by others, some of it as yet unpublished. Besides supplemental Rarámuri data from Bill Merrill and Robert Hard, I gratefully acknowledge Walter Dodd's generous sharing of information and ideas on the Guarijío.

Finally, my family continues to provide me with an inexhaustible amount of support and expertise. Thanks, then, to each one of them. To my parents, I owe special thanks for their unfailing support and nurture.

Chapter 1

Rarámuri Ethnoarchaeology and Settlement Organization

Introduction

Archaeologists currently lack methods to identify and interpret accurately certain kinds of variability in the archaeological record which represent important differences in settlement organization. This lack stems from an imperfect understanding of both subsistence-settlement systems and how these systems are manifested in the archaeological record. The ethnoarchaeological investigation of a group of mobile agriculturalists, the Rarámuri (Tarahumara) of northern Mexico, has provided the beginnings of a solution to this inadequacy.

The high degree of mobility among the Rarámuri poses a number of provocative problems for the archaeologist. An inability to interpret the Rarámuri settlement pattern in terms of their mobility could easily result in a number of wrong inferences about their culture. For example, without the historical data it is easy to imagine overestimates of Rarámuri population size based on the number of residential sites in any given community. Moreover, given the nature of assemblages found at the various residential sites, variations in assemblage size and composition could be interpreted in terms of status or wealth. Without comprehending the mobility strategies of the Rarámuri, there can be no adequate understanding of the Rarámuri system as a whole. While the Rarámuri are only one case, they stand as an important example of the inadequacy of archaeology to interpret the past using only our past methods. An ethnoarchaeological study of Rarámuri land use practices provides an opportunity to diminish our ignorance and increase our understanding of agricultural systems and the manifestations of mobility in site structural terms at a theoretical level. Studies such as the one presented here provide us the opportunity to evaluate various techniques used to extract and interpret information from the archaeological record, to assess the kinds of questions we should ask of the record, and to identify additional ways in which archaeological data indicates differences in settlement patterns that reflect variability in the overall organization of cultural systems.

The organization of subsistence-settlement systems has material consequences. If accurately identified, archaeologists can use these consequences—material remains and their arrangements in space and time—to distinguish among different subsistence-settlement systems. Some of these distinctions are apparent at a regional scale, but others are apparent only through the study of residential sites and dwellings within these sites. Important factors relating to settlement organization that are expressed in the physical organization and use of space at a site are: overall length of occupation, permanence of the occupation, disposal behaviors and site maintenance, the relative formalization and permanence of activity areas, and the arrangement of facilities.

A number of cross-cultural studies have demonstrated a strong relationship between sedentariness and agriculture. Based on a cross-cultural sample, Rafferty observed 88% of all sedentary groups included in her sample are agriculturalists, and 90% of the agricultural groups were sedentary (1985:Table 4.2 and p. 134). These data suggest an overwhelming relationship between sedentariness and agriculture. The relationship does not always imply permanence, however; the Rarámuri have been classified in the Ethnographic Atlas as sedentary (Murdock, et al. 1962-1966) although they actually have a relatively high mobility rate.

Names abound in the literature for groups with different mobility patterns; the terms most commonly used are semi-sedentary, semi-nomadic, and transhumant (Murdock 1969; Murdock and Wilson 1972; Rafferty 1985). To avoid confusion with these vaguely and variously defined terms, the very broad term "residentially mobile agriculturalists" is used in this discussion of Rarámuri settlement strategies. Residentially mobile agriculturalists are here defined as food producers whose domestic units (co-residential groups) change their residence during the year as part of their agricultural and/or seasonal round.

Cross-cultural evidence suggests that currently residentially mobile agriculturalists make up only a small part of the world's subsistence-settlement systems, but the transition from more residentially mobile to more permanently settled is a process requiring further study. Archaeologists should expect to find numerous groups of horticulturalists and agriculturalists who combined various mobility strategies with tending their crops (Flannery 1972). The conditions under which residential mobility combines favorably with agricultural practices requires further attention before we can confidently identify such mobility among agriculturalists in the archaeological record.

The distinction between mobile and sedentary agriculturalist is important, and identifying differences among mobile agriculturalists' subsistence-settlement systems is critical. It no longer benefits the archaeologist to conceptualize a simple trajectory from hunter-gatherer to agriculturalist, or from wandering nomadism to sedentarism. Rather, to understand the processes that are involved in human adaptation requires that we as archaeologists abandon these notions and frame our questions, and our attempts at understanding cultural systems, in terms of the behaviors and processes involved.

The Rarámuri data issue a provocative challenge to our assumptions about not only what we might be able to "see" archaeologically, but also why it is important. For example, when residential moves are made but site function remains the same at each residence throughout the year, what does it mean to speak of "season of occupation?" How do we incorporate a notion of seasonal change without a concomitant site-functional or subsistence change? How far apart must seasonally occupied residences be before they represent a system of "biseasonal mobility?" How shall we interpret the settlement pattern? What kinds of settlement organization would we reconstruct?

The Rarámuri

The Rarámuri, one of the largest indigenous populations in Mexico today, live primarily in the Sierra Madres, Chihuahua, Mexico (Figure 1). The more common name for the Rarámuri is Tarahumara, and most of the anthropological literature including some contemporary work use some form of this better-known name. But as they typically refer to themselves as Rarámuri, that name will be used here. There is much variability in the Rarámuri lifestyle throughout the Sierra, but many Rarámuri still live as agriculturalists whose subsistence practices and material culture have remained relatively unchanged during the past four hundred years. The Rarámuri discussed here come from in and around the community of Rejogochi. Unless otherwise stated, the following description of the Rarámuri is specific to the study area and should not be generalized to the entire Sierra.

In many ways the Rarámuri are an ideal group for a study of residential mobility among agriculturalists. The Rarámuri have two major kinds of residential mobility. They move biseasonally from warm weather to cold weather dwellings, and they move during the agricultural cycle to tend fields in other communities. In both cases, these are residential moves of some duration, where the entire household moves to another residence. Any residence may become the "main residence" during any year or for a period of several years.

Each residence is equipped to carry out all everyday, and many less common, activities. It is thus difficult to distinguish a residence serving as the main residence for any given year. Use of the residence may vary partly by location of fields and their proximity to various communities, for social, familial, and agricultural reasons. What may be a main residence in one generation, or for some years, may have a different emphasis for another generation, or in other years. While it is useful to recognize these types of residences for the discussion of the annual settlement organization, over the long term, it is not valid to perceive of these residences as main or secondary, because these distinctions are invisible archaeologically.

The mobility strategies of the Rarámuri are varied and changeable both from year to year (season to season) and household to household. This study acknowledges the high individual variation, but stresses that the implementation of various mobility options is a system-wide pattern that calls for an explanation at the level of the entire system.

Most Rarámuri own and cultivate land in a number of communities within a region of the Sierra. Land holdings tend to be small and dispersed. There are a number of reasons for this. First, soils are relatively poor and topography is extreme, thus limiting field sizes. A bilateral inheritance rule results in the division of fields between all of one"s children, which through time leads to smaller and smaller fields. Another result of the inheritance rule is that people tend to live near their relatives, siblings occupying land inherited from their parents and grandparents. Finding a spouse to whom

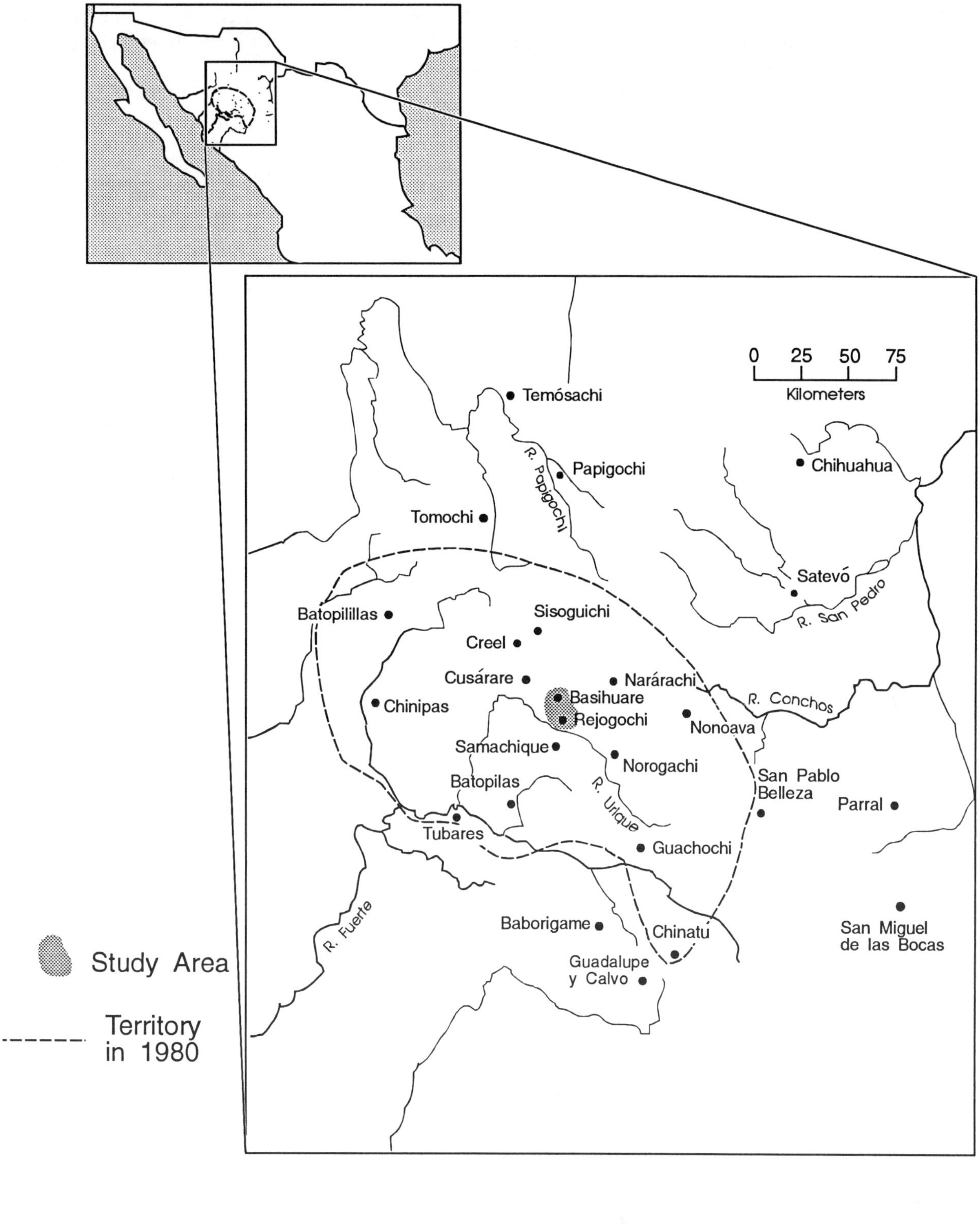

Figure 1. Map of Rarámuri homeland.

one is not related frequently requires going beyond one's natal community, resulting in households having land in at least two communities: their own, and that of their spouse.

Why the Rarámuri have a number of fields in different communities is thus an easy question to answer. However, we should also ask why the Rarámuri make residential moves when cultivating and tending these fields, rather than tending them as outfields—working in the fields during the day and returning to the residence at night, or staying at the fields for a few days while tending them and then returning to the residence at the end of that time. Obligations to friends and relatives at these other communities discourages a fieldhouse strategy. Not only do the Rarámuri use co-operative work parties to accomplish much of their agricultural labor, but the hosts of work parties also feed their assistants, and often brew a native beer typically made from corn (*tesgüino*) for them. Commitments to assist other community members in turn result in a relatively long stay in any given community.

The Rarámuri are quite flexible in their use of residential space. There is considerable overlap in the uses of interior dwelling space and exterior activity areas. The Rarámuri don't have specialized work areas. Rather, generalized activity space is used in a number of ways, and activities are accomplished anywhere within that generalized space. Facilities also retain their flexibility of placement. Maintenance of residences is frequent, but not daily. These characteristics of Rarámuri site structure are diagnostic of their subsistence-settlement organization.

Site Structure

Insofar as organizational elements are shared between different subsistence-settlement systems, the systems will have various site structural similarities, too. Site structure is defined here as the organization and use of space at a place. As used here, site structure focuses more on how living peoples define their lifespace than on the arrangements of materials archaeologists may encounter, and the latter are discussed in less detail. Site structural indicators should not be thought of in a unilineal fashion; instead, we must strive to understand how important elements of cultural systems are manifested in the ways space is arranged and the places where activities are carried out. For example, when length of occupation and dependence on stored foods are important factors for both of these groups, some site structural aspects of agriculturalists' residences are similar to elements of winter residences of hunter-gatherer groups (see O'Connell 1987). Neither the material expression of culture nor the resulting archaeological record are straightforward.

To demonstrate that site structural indicators of mobility among agriculturalists are also apparent archaeologically, the residential site structure of the Rarámuri is analyzed in the context of their subsistence-settlement system and two other groups of agriculturalists, the Guarijío and Tuxtleco. These site structural indicators can help monitor settlement organization over time and across space. Based on the analyses of this study, we can develop better methods to identify and explain variability in prehistoric settlement patterns. This study focuses on the spatial arrangements of activities and facilities at residences with the proviso that site structural analysis is only one of a number of lines of evidence that should be used in any study of the regional distribution of archaeological materials.

Residences

As used here, the term "dwelling" is limited to the physical structure that houses a combination of domestic activities—e.g., sleeping, eating, storage, socialization, and recreation. "Residence" refers to the living area that contains the dwelling. Residence is the more encompassing term, including the space occupied by the dwelling and the activity and refuse areas around it. "Residential sites" are presumed to have an archaeological manifestation; they have or had a residence, although it may no longer have material evidence.

Residences are the loci of numerous activities, performed by household members and visitors. The kinds of activities, where they are performed, and who performs them are all part of the cultural system, and for this study, the subsistence-settlement system in particular. At different kinds of residences we can expect length and intensity of occupation, food processing and preparation tasks, social interactions, maintenance behaviors, and other activities to vary. This variability is expressed in site structural terms in the arrangement of facilities and the placement of activities at the residence, and in the location and nature of refuse areas.

Although the residence is a focus for our everyday lives, much of our living goes on beyond it. Archaeologically, we break sites up into types—for example, residential, field house, special activity, and hunting stand. These categories reflect our recognition of the different places, and kinds of places, people use in carrying out their daily lives. The study of residential lifespace is informative about general adaptation and can help us to anticipate other kinds of sites. With a theory of space use, residential sites can indicate particular characteristics of subsistence-settlement systems of the past, because the activities occurring at residences are tied in with a people's overall lifeway. We can learn to link the two, and to understand the spatial patterning of structures, features, and artifact distributions at residential sites in terms of aspects of subsistence-settlement practices.

Towards A Theory of Space Use

Differing site structural properties of different kinds of subsistence-settlement organization point up important variability in the archaeological record. Maintaining a perspective of hunter-gatherer v. agriculturalist, or mobility v. sedentism, in our interpretations of the past is of limited utility. There is no doubt that archaeologists have been aware of the limitations of such a perspective for some time, and have had ethnographic evidence refuting these dichotomies. Nevertheless, few of us have modified our methods of data recovery and analysis adequately to reflect our growing understanding of cultural systems. While we have begun to incorporate this understanding into robust models of prehistoric subsistence-settlement systems, our methodology has continued to limit our understanding of the past.

Although site structural analyses have predominantly concentrated on the use of space by individuals, much work remains to be done on understanding lifespace in the context of groups of people (Figure 2). The foundation of site structural analysis links the use of space with the dimensions and mechanics of the human body itself (Binford 1983a:149-172; see also Panero and Zelnik 1979). In this context, such analysis must go beyond the basic principles of physics in explaining the placement of kinds of activity areas, the overlap of these

Figure 2. Group of people at residential site—showing various activities, age sets, etc.

areas, and the accommodation of tools, facilities, and debris within a given activity space. While a solitary person uses space in specific ways, the use of that same space by a group is influenced by more than just the combined requirements of each individual. Binford (1983a, 1978), for example, illustrates the dynamics in space use around a hearth as the number of people increases; the orientation of people sharing the same facility will change as more people require the synchronous use of that facility. O'Connell (1987) demonstrates how duration of activity area use, size and makeup of households, dependence on stored foods, and shelter requirements are important factors influencing residential site structure among hunter-gatherers.

Residences may be thought of as places where a number of individuals perform numerous and varied activities in a limited amount of space. As intensity of occupation increases, that is, as length of occupation increases, as the number and kinds of activities at the site increase, or as the number of people using that site increase (all conditions of increased sedentism, for example), a number of conflicts in space use arise that are in part resolved through site structural changes. In this study I consider a number of factors influencing residential site structure, including length and relative permanence of occupation (especially with reference to mobility strategies), subsistence activities, environment, and personnel (the co-residential group, or household).

The factors I investigate are primarily those influenced by relative intensity of occupation. As intensity of occupation increases, activity areas become more formalized and specialized; maintenance behaviors both increase and become qualitatively different. Physical consequences of these changes are the use of spatially discrete areas for different kinds of activities, under some conditions to the extent of constructing barriers to divide activity space. With increased formalization of activity space, there is also an increased investment in facilities. Facilities themselves become associated with specific places. With increasing intensity of occupation, there is a shift away from movable facilities towards fixed ones.

Thus, while the Rarámuri use their structures in a generalized fashion, the Tuxtleco and the Guarijío locate the same activities in separate areas, often setting them apart structurally, as well. Moreover, some of the facilities used by the Rarámuri tend to be less permanently fixed than those of the Guarijío or Tuxtleco. Maintenance activities and refuse areas also become delimiters of activity space.

Approaching the Archaeological Record

A theory of site structure considers not only the material expressions of the organization and use of space, but also how that expression becomes the archaeological record. Methods useful for understanding the record (ways of looking at the distribution of artifacts through time and space) are based on comprehending how materials are deposited at a place, and the subsequent taphonomic processes affecting both the materials and the setting in which they are found.

The meanings of distributions of artifacts across a landscape are being challenged as we become more aware of the ways in which materials are laid down, and the subsequent modifications of the archaeological record by people, animals, and indeed the environment itself. New approaches developed to better account for these formation processes often focus on data recovery techniques that do not mask fine-grained patterning in the distribution of artifacts on a landscape. These approaches also provide us with a means for identifying geomorphological processes on an archaeological landscape.

Many of the new techniques, while providing important settlement data, nevertheless would fail to reflect adequately the subsistence-settlement system of the Rarámuri. In conjunction with approaches to artifact distributions across a landscape, methodologies must be developed to enable us to understand the distribution of archaeological materials as produced within the context of cultural systems. Such methods should inform us about the kinds of archaeological data indicative of settlement organization and variability within that organization. The Rarámuri provide us with a very real opportunity to assess and address the inadequacies of our approaches to material remains.

Many of the standard methods used for inferring settlement organization are inadequate for dealing with the Rarámuri case (and others) primarily because they are geared toward description rather than toward asking questions about cultural processes. Such a framework lacks explanatory power. We should consider new ways of approaching the material record; how do we design our research to maximize the pertinent information we retrieve from the archaeological record if we are to ask (and answer) questions about cultural processes?

Relationships Between Artifacts and Places

With increased length of occupation, dumps become established, and maintenance behavior increases (Anderson 1982; Hitchcock 1982, 1987; O'Connell 1987; Oswald 1987; Yellen 1977). Heavily used activity areas are more frequently and more scrupulously maintained—for example, they are cleaned and swept daily. In these areas very little of the activity remains in evidence where it took place. Rather, one must look to the dumps and trash areas to find evidence of these activities (Chang 1988; Dodd 1984; Hitchcock 1982; Killion 1987; Newell 1987).

This is a qualitative difference in artifact assemblages from sites where there is little to no maintenance. The degree to which activity areas are specialized and the extent to which specialized dumps are or are not associated with specific activities or activity areas are related to the nature of the occupation of the site as well as to the kinds of activities occurring at that site. These site structural characteristics are important, but often the units of excavation traditionally used by archaeologists do not expose contiguous areas of sufficient size to show critical arrangements of features and the nature of the distribution of artifacts. A result of such an excavation strategy is that important variability in site structure and artifact assemblages literally is left buried. Baseline knowledge of how space is organized and used combined with an understanding of the nature of the relationships of artifacts and places is critical for making both informed decisions about how to recover relevant information from the archaeological record, and inferences about the systems that created it.

Sites as Resources

In some settlement systems, sites are viewed as fixed resources to which inhabitants return (for example, Binford 1979, 1982; Brooks and Yellen 1987; Chang 1988; Graham and Roberts 1986; Watson 1979). An example of conceptualizing residential sites as resources is the presence of items in active and passive storage (Binford 1979:256). With active storage, goods are put away in formal storage facilities for safekeeping, to be used at a later date. In contrast, passive storage is far less formal—items are out of the way of general traffic and activity patterns, but available to be drafted into use as needed. One form of passive storage is also called "provisional discard" (Deal 1985; Hayden and Cannon 1983; Killion 1987), where items put into refuse areas may at a later time be recalled into use. Site furniture provides another example of passive storage. Site furniture is here defined as any item left at a site, "belonging" to that site in the sense that it is left behind when the site is abandoned in anticipation of being used when the site is reoccupied (sensu Binford 1978:339; see also Brooks and Yellen 1987; Chang 1988; Schiffer 1987).

The habitation sites of residentially mobile agriculturalists share characteristics of spatial arrangement with Bushman hunting blinds (Brooks and Yellen 1987) and Eskimo fish camps (Chang 1988) in that all reflect use of the sites as resources for raw materials, tools, and shelter. In all cases, a number of pieces of site furniture remain at the residence.

One of the primary ways of distinguishing repeatedly reoccupied from continuously occupied residences may lie in accurately identifying facilities as site furniture, and understanding under what conditions these are movable, and why they become permanent fixtures in more sedentary groups. Sites that are repeatedly reused are expected to have a higher amount of site furniture than sites that are abandoned without anticipation of a return (Tomka 1989). In contrast, continuously occupied sites, while having a number of things in passive storage, are expected to have little in the way of site furniture. More permanent facilities and formalized work places are expected at permanently occupied residences.

Sites as Lifespace

> From the inception of any community, at every point in its "inhabited" phase, and during its gradual decline, people and nature in combination act as agents of disorganization. (Ascher 1969:44)

Site formation is a continual process; in the midst of pursuing activities (of living at a place) the occupants create and recreate spatial arrangements. In the process, they modify artifact associations, using their activity areas, and constructing and refurbishing their dwellings (Anderson 1982; Binford 1981; Oswald 1987; Rodman 1985; Schiffer 1972, 1987). The result of agriculturalists' occupation of a residence often is a blurring of this process. Among the Rarámuri for example, although we might distinguish between the main residence and those residences occupied during the agricultural cycle on the basis of single occupational events, in an archaeological context we are less likely to retain this ability if, over the years of occupation, site function has changed from main to secondary residence, or vice versa. Within the cultural time frame of archaeological sites, rarely can archaeologists expect to identify single episodes or events (Camilli 1983). This is well illustrated from ethnoarchaeological work with modern hunter-gatherers that documents the relative invisibility of some seasonally occupied residential camps (Ascher 1969; Binford 1977, 1982, 1983b; Brooks and Yellen 1987; Foley 1981).

While such sites are obviously an important part of a cultural system, the "quick-time" questions relating to single episodes or events per se are not necessarily those most relevant to archaeological inquiry. If ethnoarchaeologists can clarify this point, and begin to approach their data with a broader time frame, then the contributions made will be more useful to archaeology, and we will become increasingly successful in developing methods for investigating prehistoric cultural systems.

The material record observed by the ethnoarchaeologist should not be confused with the archaeological record that is recovered subsequently. Not all material consequences of a subsistence-settlement system are found archaeologically. The archaeological record may pass through various other site formation and taphonomic processes after materials are laid down as part of an occupational episode. Ethnoarchaeologists have the opportunity to consider how aspects of cultural systems are expressed in material terms, and to develop strategies of data recovery and analysis to maximize the information reflected in the record.

Procedures

The ethnoarchaeological project carried out for this study involved a 14-month study in the Rarámuri community of Rejogochi, from late January 1987 to early April 1988. During this time I mapped nine residences in Rejogochi, and a tenth in the community of Horírachi, and sketched over 35 other dwellings in Rejogochi, Bakéachi, and Basíhuare (these communities lie within the study area shown in Figure 1). Except for Residence 16, all residences portrayed as figures in this study were mapped using a Brunton transit compass and a tape measure. Residence 16 was mapped in 1983 as part of the Tarahumara Ethnoarchaeology Project (Hard 1983) using a plane table and alidade. All residences were occupied at the time of mapping except Residence 5, which had been abandoned for the winter. Most of the residences illustrated here were mapped during the initial months of my stay, in the late winter and early spring of 1987. Residences mapped during and after the growing season are portrayed slightly differently as a result of variations in seasonal uses and vegetation cover. I made observations at the residences throughout the year, with an effort to note both consistencies and changes in the daily placement of activities, objects, and features, and during special or rare events. I was personally involved with some households more than others, especially Households 1, 3, 5, 6, and 10 (see below). I visited people in these households regularly. I participated in childcare activities, food preparation and the making of corn beer, sewing, making baskets, and social and religious gatherings (work parties, death fiestas, and curings. I also observed and interviewed household members on their way of life. I conducted limited interviews on space use and compiled information on site histories. The data on site histories is somewhat uneven, and incorporates the work of Hard and Merrill (1983), personal communications with William L. Merrill, and my various interviews and conversations from my 1987/88 field season.

Summary

As subsistence-settlement organization changes, so, too, will aspects of the archaeological record that can be identified, monitored, and interpreted. The ability to observe these differences accurately requires a basic understanding of the relationships between culture and its material expression and how that material expression becomes the archaeological record. Ethnoarchaeological studies are some of the best ways of gaining such knowledge. I carried out the ethnoarchaeological investigation of Rarámuri residential site structure presented here with the goal of developing better methods for recognizing, as well as gaining a better understanding of, variations in settlement organization among agriculturalists.

Site structure, as demonstrated by this research, is a sensitive indicator of subsistence-settlement practices. A number of site structural analyses of hunter-gatherer systems have discussed the relationship between site structure and settlement organization. Comparable studies among agriculturalists are only beginning to be designed and implemented. To date there has been little done towards developing a unified theory of lifespace, and site structural studies run the risk of remaining only sophisticated pattern recognition studies. Pattern recognition studies are valuable, but they lack explanatory power. A theory of lifespace provides a context wherein archaeologists can address issues of cultural processes.

Mobile agriculturalists' subsistence-settlement systems are in many ways distinct from those of more sedentary groups. No single characteristic distinguishes between them, and such a notion would express the differences only in descriptive rather than organizational terms. A better approach is to link archaeological consequences with subsistence-settlement patterns through an understanding of the relationships between aspects of a cultural system and its material expression, and of the context in which this material expression enters the archaeological record. The goal of this study is to gain a better understanding of the Rarámuri system and prehistoric systems, and of subsistence-settlement organization in general. From this understanding, we will be better able to identify and explain variability in settlement patterns in the archaeological record.

Chapter 2
Subsistence and Settlement in Rejogochi

Introduction

The Rarámuri are subsistence agriculturalists. Principally, they grow corn, beans, and squash. They keep a variety of animals, the most important of which are sheep, goats, and oxen. They also raise pigs, chickens, and turkeys. Dogs are used to protect animals and property. The Rarámuri also have orchards, and may grow wheat, potatoes, and some vegetables. They collect wild resources as foods, medicines, and trade items. The Rarámuri participate to a limited extent in the Mexican economy. Although not as isolated from outsiders as in the past, ethnographic accounts written in the late part of the last century and the early part of this one still describe much of the social and religious organization, as well as the subsistence-settlement system and material culture of the modern-day Rarámuri.

The community of Rejogochi lies in the transition zone between the deep, narrow canyons, called barrancas, and the broad, flat uplands of the Sierra Tarahumara. Rejogochi lies in a broad valley bottom about 2000 m above sea level, with mountains rising another 200 m from the valley floor (Figure 3). Residences extend for approximately 5 km along the valley, through which runs a perennial stream. Warm weather residences are generally on or near the valley floor, while winter residences lie on small knolls or in rockshelters on the mountain slopes.

Rejogochi Subsistence Practices

Annual Subsistence Round

In Rejogochi, farming activities begin in late winter when the Rarámuri plow fields in preparation for planting maize, and cut trees for the slash and burn fields in which they will grow beans. They cultivate today with oxen-drawn plows (Figure 4). These activities frequently are co-operative, with households joining together to assist in preparing their own and each others' lands (see below).

Also at this time, vegetation around fields is burned. The skies are hazy with smoke at these times and again just before the rainy season begins when the bean fields are burned. Lumholtz (1902) reported that Rarámuri did this to bring rain. Rarámuri in Rejogochi say they burn because the fields are dry, and can be burned, and to eliminate vegetation that would get caught in their plows (W. Merrill, personal communication, 1992). I was also told burning prevents weeds from encroaching and discourages animals from entering the fields to forage. Cooperative labor groups carry manure down from the rockshelters and barns where animals were corralled during winter nights down to fertilize the fields.

As the weather begins to warm, households from Rejogochi move from their cold weather residences to their warm weather residences. A number of factors are involved in the decision to move, including general climatic conditions, the location of the warm weather residence and the point at which it receives early morning sun. Most cold weather residences are located close to seasonal seeps and springs. These water sources are dependent in large part on annual precipitation, especially that accumulated during the winter months. When seeps and springs begin to dry up, generally in February and March, households move to residences located closer to other water resources—most commonly near to the perennial stream running through the valley.

Figure 5 graphs precipitation in Rejogochi for 1978 and portions of 1977 and 1979. Precipitation comes mainly in the form of rain, but also as snow, sleet, and hail. The success of the corn crop depends on sufficient moisture in the soil from winter precipitation, especially snow, for the seed to germinate prior to the beginnings of the summer rains.

The Rarámuri plant corn and squash in April and May. Planting is one of the busiest times of the year; only during planting and harvesting did people consistently phrase their planning in terms of these activities (e.g., "after the planting"). The

Rarámuri who practice Catholicism are called *pagótame*, meaning baptized (see Bennett and Zingg 1935:182, 319-325; Kennedy 1978:127-156; Lumholtz 1902:295-296; Merrill 1988a:46-52 for discussions on Rarámuri religion and the differences between the Christian and non-Christian Rarámuri in the Sierra). Easter is the first Sunday following the first full moon after the vernal equinox. Holy Week, the week culminating on Easter Sunday, is the most important religious event celebrated by the Rarámuri, and usually falls around this time. A number of preparations before and during Holy Week require the Rarámuri to reside in the Basíhuare center where the mission church is. Consequently, planting activities are arranged around the Holy Week celebrations.

Besides tending to their fields in one community, a household may also plant in other communities. In many cases, the extreme topographical variation of the Sierra provides enough of a difference in temperatures, precipitation, and soil moisture regimes so that planting and harvesting can be done sequentially from each of their fields in different communities. Thus, communities even within a one hour walk of Rejogochi might vary a week or two from Rejogochi's plowing and planting sequence.

This variation has important implications for the scheduling of a household's agricultural activities when it cultivates land in more than one community. Besides the relevance of environmental factors, the coordination and scheduling of work parties also plays an important role in the timing of agricultural activities. While households do some of their own planting, it is also a time of joining and sponsoring many work parties, with people coordinating their planting schedules and drinking parties that accompany the work.

Shortly after the corn is planted and has sprouted, the Rarámuri begin collecting wild greens from the fields. Greens are an important part of the Rarámuri diet, being one of their few sources

Figure 3. The lower valley of Rejogochi. This photograph was taken facing downstream (southwest), near the mid-point of the valley. An entrenched perennial stream runs through the valley, and a logging road roughly parallels the stream. Houses and fields occur on both sides of the stream; many of the houses are located near the base of the slopes rising from the valley floor.

of vitamin C, as well as providing other vitamins and nutrients (Bye 1976, 1979, 1981; see also Urban 1975). While some types of greens occur and are collected beyond cultivated fields, the majority that the Rarámuri use depend on an anthropogenic environment of cleared and disturbed land in which to grow. Harvesting greens at a tender age is an effective weeding strategy in the corn fields, and provides an additional harvest from their poor soils, thus increasing the nutritional productivity of the fields (Bye 1976). This harvest of feral greens continues into July and occurs again in the late fall. The Rarámuri gather both indigenous and introduced species (Bye 1979, 1981; Pennington 1963:124-129).

Several weeks after corn is planted, family members reseed those areas where no corn has sprouted. When the seedlings are no more than 10 cm tall, hoeing begins. Hoeing and weeding are the major activities of the early summer days, and while some work parties are arranged, it is often the household or family members that perform these chores. Throughout the agricultural cycle people carry out tasks at their various fields, traveling to other communities in order to tend secondary fields and to participate in work parties of their friends and relatives in those areas, frequently staying between a few days and several weeks before returning to their primary residences.

While the corn is less than 20 cm tall, and prior to the rainy season, the Rarámuri burn off the vegetation from their slash and burn fields. Nutrients from the ash make these fields fertile and productive. Beans are planted about the time of the first rains of the year. Informants from Rejogochi stated emphatically that slash and burn plots are planted once and never again, but this may not be the case throughout the Sierra (e.g., Bennett and Zingg 1935:36; Lumholtz 1894). Swidden fields are generally recognized as being both productive and maintenance free (requiring neither additional fertilizer nor weeding), although

Figure 4. A work party plows a field using traditional Spanish plows and teams of oxen. A household invites family and friends to help with this and other agricultural activities. In turn, the household will help its neighbors to prepare their fields. The elevated structure to the right of the field is a chicken coop; nearby are several fruit trees.

insects and other pests can destroy the crop. Pennington (1963:49) suggests the bean crops from slash and burn fields do not contribute a large portion to the Rarámuri diet, but there is a great deal of regional variation. In the study area, many planted bean fields; however, it is my impression only the very wealthiest (and often small) households had sufficient supplies of home-grown beans to last the entire year, while the majority of households depended heavily on purchased beans to supplement their own crop.

The summer rains officially begin on June 23rd, when the corn is about 20-30 cm high. Rain comes almost daily and can be torrential. Paths and traffic patterns change as the streams swell and fording areas are drowned. Seasonal streams run once again, and in Rejogochi some Rarámuri use these water sources rather than the perennial stream. The runoff also fills large natural pot holes, or tanks, in seasonal stream beds that act as water reservoirs for the rest of the year. Seeps excavated near the streams provide additional sources of drinking water.

Throughout the growing season wild animals and birds threaten crops; crows are carefully watched and children sent running to the fields when crows circled overhead. From the time of planting until harvest, domestic animals are watched carefully and controlled. Prior to planting, sheep, goats, pigs, and cattle are allowed to roam fairly freely; cattle and pigs are left unattended, and sheep and goats watched casually. Their manure is vital to continued field fertility. During the growing season pigs are penned. Sheep, cattle, and goats are taken out daily to forage, and are closely supervised. Fodder, put away at the end of the previous summer, is used to supplement the stocks' diet during early spring when forage is scarce, or at times in the summer when a household is too busy (for example, when preparing corn beer or planting) to go out with the animals.

During warm weather, animals are corralled at night directly on a field in fallow. Rarámuri will move the corral every couple of weeks, thus ensuring that most or all of the field is fertilized. Fallow portions of fields in production are fertilized in the same way. Prior to the introduction of chemical fertilizers (which are still cost-prohibitive) manure was the only fertilizer available. In other parts of the Sierra, Pennington (1963) reports the use of bat guano as fertilizer. Lumholtz observed that the Rarámuri also move their homes periodically in order to plant this more organically rich soil (1902:171).

The time of greatest food scarcity is that part of the summer when the stored food supplies are depleted, and before the green corn is ready. Women and children gather wild foods from the surrounding area. They harvest both wild and domestic greens, like the leaves and flowers of squash, from nearby fields and gardens to supplement the basic diet of corn and beans. Many Rarámuri also depend on money earned during the year to purchase additional foods, especially at this time.

The Rarámuri fish throughout much of the year, excepting the rainy season when streams are turbulent. Between March and June when agricultural stores are depleted and wild plants are as yet

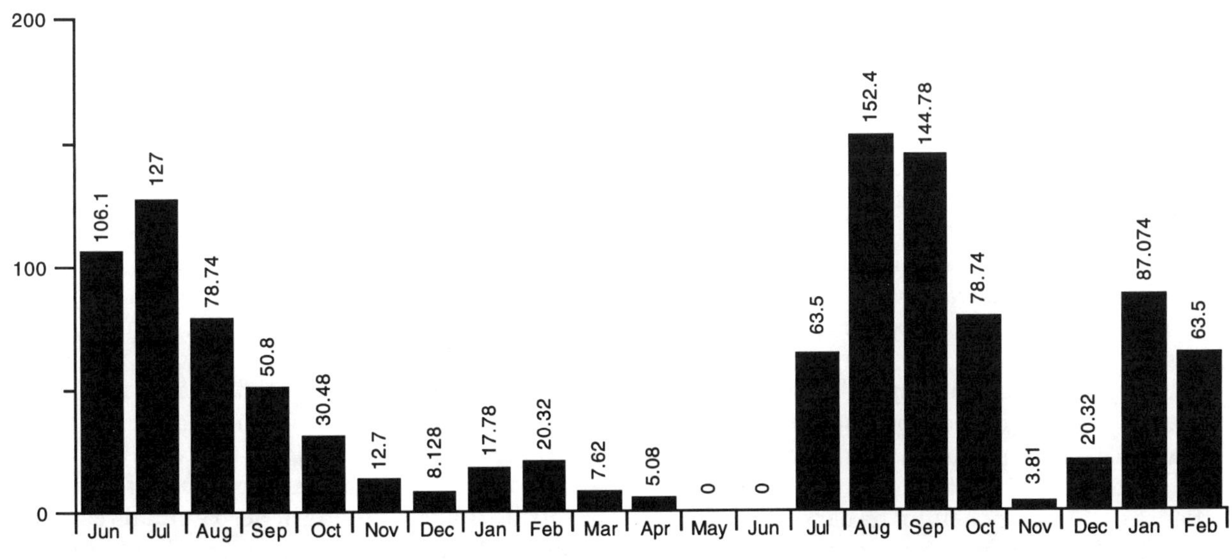

Figure 5. Precipitation in Rejogochi 1977–1979 (in mm), from W. Merrill, 1977–1979 field data.

unavailable, fishing from nearby areas can be an important supplement to other subsistence activities. Households often spend a day out fishing and picnicking, although it is often the men and boys who do most of the actual fishing. In Rejogochi, the Rarámuri trap numerous kinds of rodents and other small game as additional food sources. The role small game plays in the Rarámuri diet is not well known. Such animals probably are an important source of meat for them. The population density of large and medium-size game is low throughout much of the Sierra. In the study area such animals rarely are observed.

The Rarámuri begin eating the fruit from their fig, apple, and peach trees while it is still green. The trees are typically near residences, including some winter homes. Rejogochi has one orchard, planted between 1983 and 1987 with the help of the Jesuit mission in Creel. Residences (including permanently abandoned residences, obviously) may have one or several trees associated with them.

The rains temporarily cease towards the end of August, then return again for part of the month of September. This last period of rain is important for producing a good bean crop (Bye 1976; Pennington 1963). As the Sierra begins to dry out, the corn stalks turn brown and the ears begin to dry. Grasses used as catalysts in fermenting homemade beer ripen at this time and women harvest the seed. The Rarámuri continue harvesting and eating greens they plant in the manured areas of corrals (*Brassica* sp. and *Lepidium* sp.; see Bennett and Zingg 1935:39-40, 143; Bye 1979, 1981; Pennington 1963:67). Some green corn is eaten, either roasted or ground and made into tortillas. Even at this time, however, the principal foods are made of dried corn from the previous harvest.

In early autumn, Rarámuri men and boys cut and collect corn stalks, tying them into bundles. Households store this fodder in trees or sell it to local mestizos. Apples and peaches are harvested. The apples are stored to be eaten through the winter and early spring. Fruit is also sold in Creel, or to mestizo traders who come into the valley. The nuns who work at a health clinic in Creel will also exchange potatoes and maize for some of this fruit harvest.

Everyone in Rejogochi spends most of September and October harvesting their crops—fruits, squashes, beans, and maize. This is a time of plenty, and people give a portion of these foods to each other in exchange for help in the harvesting. Guests are offered a few ears of corn to take with them at the end of a visit. This is a busy time, as everyone tries to finish harvesting and storing crops. Ears of maize are picked off the stalks in the field and then brought back to the residence to finish drying (Figure 6). After the corn is dry, it is stored without shelling. Some ears may be shelled and the corn spread to dry in the sun for a day or two if the storage structure is full, or perhaps if the previous year's harvest is nearly gone and the household anticipates needing dried maize in the near future. Sometimes drying and storage structures are placed in fields, but more typically at the residence. Households with fields outside Rejogochi may be engaged in harvesting activities even into November. The amount of time spent at these tasks varies with the timing of the agricultural cycle in other communities where they have fields or are part of a cooperative labor network. Following the harvest, and with the onslaught of cooler temperatures, households again move to their cold weather residences.

Domestic Animals

Domestic animals play an important role in the Rarámuri way of life in Rejogochi. In the seventeenth century, the Spaniards introduced cattle, sheep, and goats into the Sierra. The Rarámuri quickly incorporated the animals into their economy. They especially valued the manure of these animals as fertilizer (Figure 7). In combination with the introduced metal axe for clearing land, cattle and the plow allowed the Rarámuri to cultivate larger fields more efficiently. Sheep were also prized for their wool, from which both clothing and blankets were made. These domestic animals continue to be important in contemporary Rarámuri society for principally the same reasons. In Rejogochi, the Rarámuri rarely milk their animals; when they do so, the milk is made into cheese, rather than drunk. The Rarámuri kill and eat cattle, goats, and sheep very rarely; usually for some religious celebration.

Few of the Rarámuri near Rejogochi have horses. Dogs are important for guarding animals while they graze and at the residence. They are also used in hunting wild game. Chickens and turkeys are both kept as sources of meat and eggs. Pigs are kept as a meat (and fat) source.

The number of domestic animals varies considerably among households. Some will have none, while others have large herds of sheep, goats, and/or cattle. The care-taking of herd animals can be a rather arduous task, especially during the growing season because they must be taken out to forage daily. A household can lend some or all of its herd to one that doesn't have these animals. In exchange for the care-taking of the animals, the household borrowing the animals is able to manure its own fields, thus increasing crop yields.

Participation in the Cash Economy

In the study area, the Rarámuri continue to be subsistence agriculturalists, relatively independent from the Mexican economy. However, wheat flour, sugar, coffee, salt, and lime (used to process corn) are all purchased staple items the Rarámuri consider necessary to their diet. Cloth for clothing, and metal tools and utensils are also items that must be purchased with cash. Nearly all households do something in order to acquire money. Animals sold to mestizos are generally paid for in cash. Between the Rarámuri themselves, animals and land are usually traded or exchanged for comparable goods, but may also be purchased for cash. Some Rejogochi households depend on wage labor to supplement their subsistence agriculture, deliberately underproducing corn and beans and instead buying more of their food (W. Merrill, personal communication, 1989).

Work at the *ejido* sawmill, work on the roads, and occasional work in the towns of Creel and Samachique are ways in which Rarámuri men can earn money. Men go to Cuahuatemoc, Chihuahua City, and beyond, for work. Beginning at about age 15, boys will spend most of the next several years working, often staying away from the community for extended periods to acquire money before marrying. Women generally make various craft goods to sell to tourists in town for cash. Rather than selling directly to strangers, the Rarámuri prefer to sell their hand-made crafts to traders or shop owners in Creel or other towns. These merchants then sell to the tourists. Home-grown fruits and vegetables are sold to mestizos in town or traveling traders, and the Rarámuri also collect wild foods

Figure 6. A couple harvesting ears of dried corn from their field. They will carry the corn to their residence using the bags in the foreground. There, they will allow the corn to dry completely, if necessary, and then put it in their storage structure.

(for example, greens and mushrooms) and medicinal plants, taking them as far as the markets in Chihuahua City to sell (Bye 1976).

Not all interactions with mestizos include the exchange of goods for cash. The role of mestizos in the network of traders who trade goods from one part of the Sierra to another has been briefly mentioned by early ethnographers and continues to the present (Bennett and Zingg 1935; Lumholtz 1902; Pennington 1963). To date, very little is known of these trading networks or interactions wherein specialized goods and foods move between the barrancas and the uplands.

Land Use Strategies in Rejogochi

Biseasonal Mobility

In the study area, the Rarámuri make their seasonal move to cold weather residences around November, when temperatures drop, the winds become bitter cold, and chances of precipitation increase. Temperatures in Rejogochi are relatively mild, but with seasonal extremes of hot and cold. Figure 8 shows the mean monthly highs and lows for 1978 and portions of 1977 and 1979 (W. Merrill, unpublished fieldnotes, 1977-79). Throughout most of the year, the temperatures during the day range between 15 and 20 degrees C, and it is reasonably comfortable to work out of doors.

Many Rarámuri move to cold weather dwellings to be closer to water and fuel resources, and to obtain greater protection from wind, precipitation, and cool temperatures for themselves and their animals. In Rejogochi, the move from the warm to cold weather dwellings is often associated with moving domestic animals from open air corrals to warmer, more protected, shelters (Hard and Merrill 1992). Frequently, although not always, these cold weather dwellings are rockshelters (Figure 9). In

Figure 7. Corral in a recently harvested field. The cattle are penned overnight in the corral, thus manuring this part of the field. After a couple of weeks, the corral will be moved to a new section of the field, and the process repeated. Goat and sheep corrals are usually a little smaller than the one shown here.

other parts of the Sierra Tarahumara, true caves are also inhabited. The Rarámuri modify rockshelters to provide not only comfortable living space for themselves, but also enclosed corrals for their animals. Ownership of animals is also seen as a critical factor in biseasonal moves in these areas:

> The possession of livestock, particularly goats and sheep, imposes on their Tarahumara owners a seasonal pattern of nomadism from which those without such animals are exempt.... Often, in preparation for the yearly moves [to spend the winter in] the barrancas, a man has an extra field in the canyon bottom which he plants and cultivates so that it will yield its crop in time for the arrival of the migrants. (Champion 1962:105-106)

In Rejogochi, if a household has an old or unoccupied house available it may use it as a barn for animals, in which case either the household moves to this location (where presumably there is also a dwelling for it), or it moves the barn to a suitable and convenient cold-weather residence.

Above, Champion characterizes the use of fields in the barrancas as a specific winter subsistence strategy. The use of these crops as the primary food source while away from the main community can be likened to use of agricultural produce at residences occupied during the agricultural cycle. Winter sites used by residents of Rejogochi are sufficiently close to the main residence so that access to the food storage locus is convenient. In Rejogochi, corn is stored year-round at the main residence and it is simply a matter of making a short trip (perhaps 10-30 minutes' walk) to retrieve several days' or a week's worth of corn.

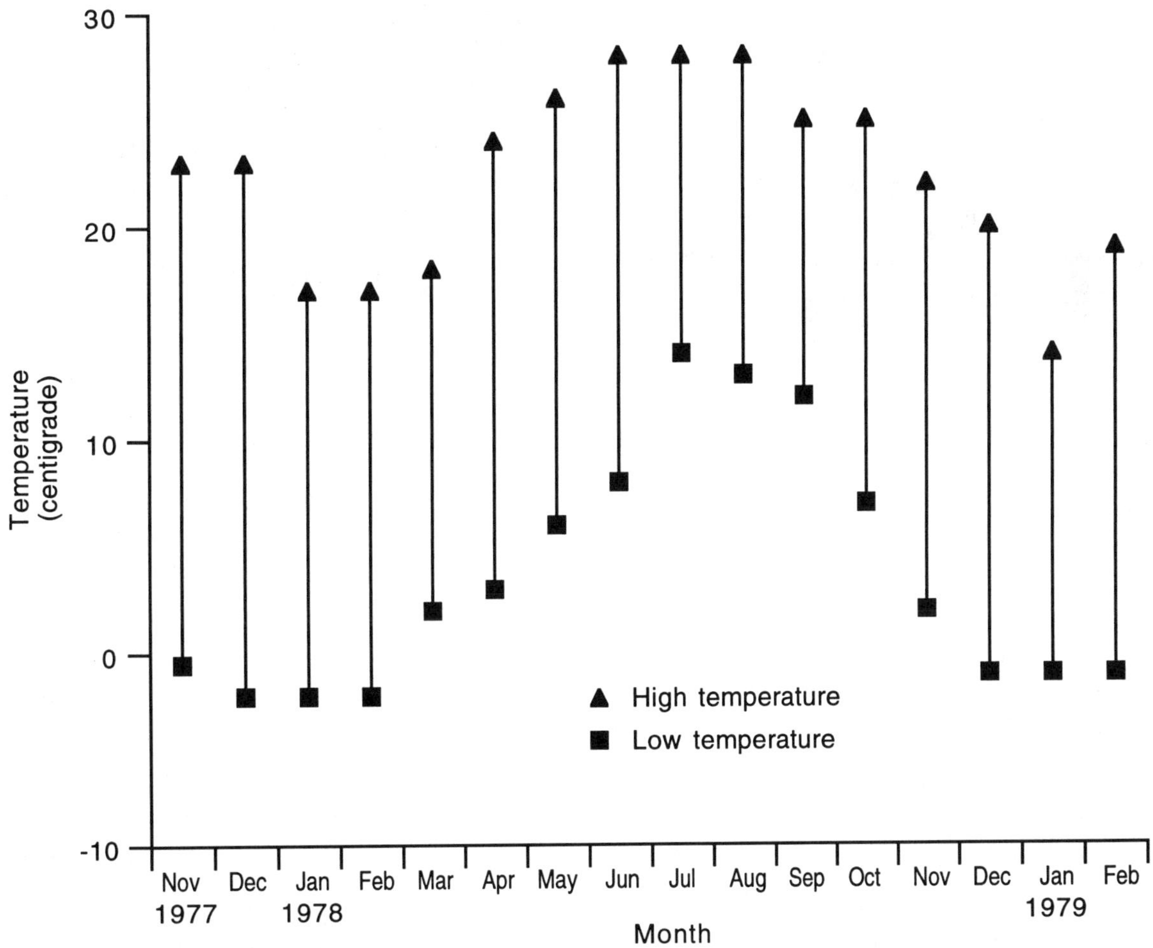

Figure 8. High and low temperatures in Rejogochi 1977–1979 (in degrees centigrade), from W. Merrill, 1977–1979 field data.

Mobility During the Agricultural Cycle

In the case of seasonal moves, the household changes its residences from the warm to the cold weather dwelling. But the household may also cultivate fields away from its main residence and often beyond the boundaries of the community. At these times, the household generally occupies a residence near its fields. Land use and mobility strategies vary widely throughout the Sierra, but generally the Rarámuri have a strategy of dispersed field cultivation (see below). As a result of this settlement strategy, the distinctions between the main and other residences are often blurred—the "main" residence one year may not be occupied as such in the subsequent year. The potential for each residence and community to become the primary home base is a factor crucial to understanding Rarámuri settlement organization. That is, we cannot anticipate which settlement will house the main residence for a given family in a given year, but we can assume that each residence will be established in anticipation of occupancy someday.

The distribution of a household's fields is in part a function of the scarcity of arable land. Hard and Merrill (1992:605-607) estimate 40 hectares of arable land in Rejogochi (Figure 10, see also Figure 3). Field sizes in this community range between 0.24 and 2.91 hectares. About one half of the households who plant in Rejogochi also cultivate fields in other locales. In much of the Sierra, the topography is extreme, and only small areas are suitable for cultivation. Such terrain requires that a number of fields be cultivated simultaneously in order to obtain sufficient amounts of subsistence foods. Champion characterizes Rarámuri field sizes

Figure 9. Rockshelter used as a winter residence. One of the two separate dwelling structures is shown in this photograph. The dwelling has a log roof extending out beyond the dripline of the rockshelter. A number of containers and utensils are sitting on the roof. The area in front of the structure is swept clean. The woodpile lies at the edge of the maintained area. A newly constructed plow leans against a tree stump. (Originally published in Graham 1993:28, Figure 3.3.)

as ranging from a quarter acre (0.10 ha) to six acres (2.4 ha) (Champion 1962:100). According to Champion, each household cultivates between two and five fields, depending on the size of each field, only one or two of a household's fields will be located in its main community. Bennett and Zingg write:

> Scattered throughout the mountains are many such family properties, composed of small fields of corn and a plank shanty to one side. Here the family lives, takes care of the field, and herds the animals.... A family has perhaps two or three of these cornfields located in different sections. Near each, a house is built, and the members of the group move successively from one farm to the other as their work dictates. (Bennett and Zingg 1935:183)

As stated above, the timing of agricultural activities may spread over several weeks in different communities, allowing for successive movements to tend dispersed fields. To check on the crops or to tend a field, one or more household members (usually men or older boys) may take a day or part of a day to walk to a nearby community. Frequently, however, the entire household will make extended visits to these secondary communities. These residential moves take place if the work will take more than several days to complete or if the household is committed to helping friends or relatives in that community with agricultural (or other) activities.

Fields in the Sierra generally have nutrient-poor soils (Bye 1976) requiring frequent and long fallow periods or the supplemental use of fertilizers. Where fertilizers are not available, land ownership is probably more extensive to allow for more land to remain in fallow while other fields are cultivated (Champion 1962:104). The manure supplied by sheep and goats, and cattle to a lesser extent, continues to be the major source of fertilizer in the study area. In this way, the presence of domestic animals influences the amount and distribution of agricultural land a household may have

Figure 10. Rejogochi has a broad valley bottom, compared to other valleys in this area. Rarámuri households plant extensively on the valley floor, as well as in swidden fields on the mountain sides, as well. This photograph was taken facing upstream (east), near the mid-point of the valley.

in production. As the Jesuit mission makes available commercial fertilizers at lower-than-retail prices, this system is more likely to change.

Anthropologists have also considered the distribution of fields a risk-reducing strategy (Hard and Merrill 1992:609-613; Merrill 1981:54-55; Sheridan and Naylor 1979:72). Given that the distribution and poor quality of arable land inhibits growing large quantities of foods in single fields and that insects, drought, and hail occasionally destroy crops, the dispersal of agricultural lands throughout a region minimizes a household's chances of losing all of its produce. This is not an overt strategy in the study area, however. Merrill points out the advantages of having fields in diverse terrains and environments, but notes the Rarámuri also employ a number of strategies to consolidate their land holdings. These include buying, selling, and trading land, as well as entering into marriages not strictly in keeping with marriage rules when it is expedient (Merrill 1981:54-56). Such attempts to consolidate fields occur throughout the Sierra. Kennedy (1978) notes that a household may lose rights to plots of land eventually when it is not able to tend all the fields because their locations are too scattered.

Ethnographers agree the distribution of agricultural and grazing resources has been important in determining Rarámuri settlement sizes and locations (Bennett and Zingg 1935; Kennedy 1978; Merrill 1981, 1988a, Pennington 1963, 1983). But whereas the size and number of fields is attributed to the scarcity and relative productivity of arable land, a number of other factors contribute to the character of Rarámuri settlement organization. There are a number of ways in which farmers can tend dispersed fields; among the Rarámuri, one of the most likely choices is to make a residential move to the communities where these fields are located. That the Rarámuri choose to make these residential moves, rather than daily field trips from more permanently occupied residences has important implications for the settlement pattern we will see archaeologically. A discussion of some of the other factors influencing Rarámuri settlement organization follows.

Bilateral Inheritance and Marriage Rules

One major factor in the distribution of a household's residences throughout several communities is the bilateral inheritance rule. This rule entitles all children to equal shares of each parents' property, thus ensuring that the cultivated lands of any household are spread throughout a larger region. The consequences of this inheritance pattern are twofold (Merrill 1981:53-55). First, as a result of the distribution of land in such a system, a parcel of land is divided up between descendants, and people who are related live near one another. Secondly, because ideally the Rarámuri do not marry anyone with whom they can trace kinship ties, marriages often take place between individuals from different communities, sometimes at some distance from each other (Merrill 1981; Pennington 1963:17).

Isolationism, Tenacity, and Survival

> From infancy they are brought up to be mountain vagabonds with no training in either morals or proper conduct.... These Indians have spent all their lives scattered about in the deepest canyons.... There is no doubt that it is a service to Both Majesties [i.e., God and the King] as well as for the good of the souls of these wretched Indians when the ministers of the King Our Lord...(God keep him!) come to take them from the canyons, forcing them to build homes and huts in the pueblos and to be reduced and live there permanently.... (Sheridan and Naylor 1979:81)

These words of Father Abee, written in 1744, reflect some early perceptions of the Rarámuri by Europeans. Throughout the New World, the Spanish had a program of congregation, or reduction, whereby indigenous peoples were consolidated into centralized settlements rather than left in the traditional communities occupied before contact (Spicer 1962). This program radically changed the settlement systems of a number of groups throughout New Spain, but nowhere was the program less successful than among the Rarámuri (Sheridan and Naylor 1979; Spicer 1962). One source of frustration for the Spaniards in their program of congregation was that Christianized Rarámuri frequently retreated from the mission towns back into their "scattered" settlements. The Rarámuri withdrew when the missionaries' or soldiers' efforts of civilization became too zealous, or Spanish and Christian customs too burdensome. They also left simply so they could tend to their families and crops. The missionaries saw the refusal of the Rarámuri to live in towns and villages as irrational and a rejection of the Christian doctrine and lifestyle.

Ethnographers have incorporated into their own writings the explanations the early Jesuit missionaries offered for the failure of congregation program. Bennett and Zingg, for example, suggest that the Rarámuri live in small family units alongside their fields not only because of the distribution of arable land, but also because the Rarámuri are isolationists (1935:183-184). They describe the unoccupied houses in the community of Samachique in the following way: "The desertion of these houses vividly exemplifies the failure of this congregation, not because the idea was unsound, but because the participants were unwilling...." (Bennett and Zingg 1935:184).

Bennett and Zingg (1935), Kennedy (1978), and Pennington (1963) all suggest the degree of mobility in the Rarámuri settlement organization is an important part of the reason this system has been perpetuated; because the Rarámuri move both during the agricultural cycle and biseasonally, they are less inclined in general to settle down. The ethnographies are not clear on whether residential moves to tend agricultural fields occurred within an annual cycle, or if these moves were each of a few years' duration, however. Sheridan and Naylor (1979:104) suggest the goods and domesticates introduced by the Spaniards served to reinforce Rarámuri settlement organization rather than to redirect it into a more populated, permanently settled village pattern. As a result, the lifestyle of these mobile agriculturalists has persisted.

Labor Organization

Small field size, necessity of fallow periods, inheritance rules, and at times, the location of the winter residence, are all reasons for the distribution of fields throughout a number of different communities. Oftentimes, the consolidation of agricultural land is an on-going process for many Rarámuri, too. Nevertheless, an important question regarding Rarámuri land use strategies has still not been addressed: Why do the Rarámuri occupy one residence after another, rather than remaining in their main community, simply setting up fieldhouses (smaller, more impermanent, structures) and venturing out for no more than an afternoon, or a day or two to tend fields away from the main community? It is suggested here the Rarámuri maintain residences in other communities in part because of their organization of labor and their social ties.

Each Rarámuri household is part of an established network whose members pool their labor in order to accomplish the task at hand (see Figure 4). It is possible that the Rarámuri would find themselves unable to extend their time and labor over the distances and numbers of fields they cultivate without the cooperative help of others (Bennett and Zingg 1935:199-200; Kennedy 1978:70-79; but see Champion 1962:127). No study has collected data to assess the amount of land a household has as compared with the amount of land that household works through its network ties, but such a study would allow us to better evaluate the relative importance of labor pooling for completing subsistence and other tasks.

The homemade beer called *batari* or *sugi* by the Rarámuri is known as *tesgüino* by the Spanish. The drink is often made of fermented corn, ground and boiled with various herbs (Pennington 1963:149-157). This beer is a central focus in the life of the Rarámuri. Discussions of beer and drinking pervade conversations—people are constantly speaking of what happened at the last work party or death fiesta where there was drinking: the strength of the beer, what people did while they were intoxicated, and plans for future drinking parties. Merrill estimated drinking was the topic of at least half of the conversations he held with Rarámuri during the first two years of his field work (Merrill 1981:198).

The making of corn beer is at least a ten-day process (Kennedy 1963; Merrill 1978; Pennington 1963). Frequently a household makes the beer while occupying a residence outside its main community. This residence must have the necessary supplies—cookware, grinding stones, dried maize, and so on—in order to carry out the steps of brewing beer. Kennedy (1963:634-635) estimates the Rarámuri spend one third of their lives preparing, drinking, and recovering from beer.

Whereas the Rarámuri of Rejogochi occupy a transitional zone between the uplands and the barrancas, Kennedy worked exclusively with non-Christian (*gentile*) Rarámuri in the barrancas under different topographical, environmental, and social conditions. Nevertheless, Kennedy's discussion of the nature of Rarámuri cooperative labor and the concept of the *tesgüino* network has relevance for the inhabitants of Rejogochi. Those individuals who participate in each others' drinking and work parties form what Kennedy labeled as the "*tesgüino* network."

> Since the fields of all Tarahumara families require the same kind and roughly the same amount of cooperative help, it is highly expedient that the place of work rotate from one rancho to another.... Economic requirements of daily

> life, seasonal moving, and often necessary movement to an in-law's rancho all act dynamically to weaken territorial bonds.... Under the conditions described here the result of environmental conditions and human propensities and limitations is the *tesgüino* network, a centrifugal household centered social system composed of many overlapping dyadic ties. (Kennedy 1978:213-214)

The *tesgüino* network has resulted from the dispersed nature of Rarámuri communities, combined with the necessity for greater-than-family size work parties. When a household needs additional help in completing a task, it enlists the aid of friends and relatives. Payment for these services are meals while working, and the reciprocation of help in their own times of need. Often *tesgüino* is also prepared for the work party. These work parties are common occurrences at times of planting, plowing, weeding, harvesting, clearing slash and burn plots, and house-building. Assistance comes from friends and relatives in the various communities where the household resides, or with whom it has relationships.

The *tesgüino* network is one way the Rarámuri maintain economic and social relationships with people in other communities where they not only tend their land, but also lend assistance to members of that community. As with the inheritance rule, the *tesgüino* network reinforces the ties of a Rarámuri to communities and households beyond the primary community.

Not all households draw on the same pool of labor resources when requesting aid. The practice of inviting workers from the various communities a household occupies serves to extend the network out over a broad area. Given such a network, an absence of centralization becomes important for maintaining an even spread of effort and obligations. That is, rather than some households or communities lying on the periphery of a centrally based organization, the organization itself shifts.

Because relationships in any community involve helping others as well as being helped, short-term trips to fields are not always sufficient or appropriate to carrying out all the tasks to be performed in the community of the distant field. Members of a household may work in their own fields one day, but on the following day they may be asked to join a neighbor's work party. Fieldhouses are inadequate given shelter needs, storage requirements, and the activity structure of these longer periods of occupation. The occupation of residences for some extended lengths of time, and the use of residences per se instead of field houses relates to the relative interdependence of households within communities and across community boundaries.

Moves to the Religious-Administrative Center

One other significant type of residential move is that made to the religious-administrative center for periods of a few days to a few weeks during times of religious celebrations. Holy Week (*Semana Santa*) is the most important religious holiday for the Christian Rarámuri (Kennedy and Lopez 1981; Merrill n.d.). The Fiesta of Our Lady of Guadalupe (December 12) and the Fiesta for the Immaculate Conception (December 8) are also important holy days in the study area. For these occasions, officials and *fiesteros* (the hosts of these fiestas) are responsible for making corn beer and hosting fiestas in Basíhuare, the *ejido* center (these celebrations are discussed in more detail in Merrill 1981, 1983, n.d.).

While officials and *fiesteros* occupy various residences in Basíhuare for the duration of the religious occasions, the remainder of the participants also must find lodging for the time they are in Basíhuare. A number of rockshelters and abandoned houses in Basíhuare are occupied only at these times. Bennett and Zingg similarly report on houses near the church in Samachique that had been abandoned for several generations, but were maintained and used (occupied) during religious celebrations (Bennett and Zingg 1935). These brief moves for religious obligations present an important and interesting dimension to Rarámuri settlement organization; however, no site structural analyses of these residences were carried out at this time.

Changing Residences—Households in Rejogochi

The Rarámuri Domestic Unit

Household structure plays an important role in determining site structure at the residence. The household is a co-residential group made up of individuals who share residential space (see Bender 1967; Netting et al. 1984a, 1984b; and Yanagisako 1979 for in-depth discussions and further references on the terms and concepts of the household, family, domestic group, and co-residential group). The basic economic unit of the Rarámuri is the nuclear family, consisting of a married couple and their children (Bennett and Zingg 1935; Kennedy 1978; Lumholtz 1902; Merrill 1983; Pennington 1963, 1983). The household itself does not necessarily break along these same lines, however. Out of 31 households, in Rejogochi, roughly 16 (51%) are nuclear families, and 15 (49%) are extended families. During any given year, and over a period of years, households demonstrate great flexibility as nuclear families join, break from one extended family to join with another, or become independent. But members of nuclear families always travel together.

After a couple marry, they usually share both their parents' residences, moving between them throughout the year. Often, they are given land at the time of their marriage and also make residential moves to tend these fields as well as to assist friends and family in their agricultural duties. After a number of years, the couple begins constructing a dwelling of their own. This is sometimes built at the residence of one set of parents or the other, but occasionally the couple will establish a separate residence at this time—for example, if there are domestic tensions between the generations. As parents or relatives become enfeebled or widowed, residence patterns may change, and these family members may join another household for one or more seasons of the year. Residential moves of households may be further affected if the spouse or relative is too ill or weak to travel to another community.

Household Mobility Strategies

> The Tarahumare [sic] never lives all his life in one house or cave; nor will he, on the other hand, leave it forever. He rarely stays away from it for more than two or three years. A family, after inhabiting a house for a time, may suddenly decide to move it, even if it is built of stone. The reason is not always easy to tell. One man moved his house because he found that the sun did not strike it enough...and many...superstitious apprehensions of one kind or another may thus influence the people. Very often a man moves for the sake of benefitting the land, and after tearing down his house he immediately plants corn on the spot on which the house stood. A family may thus change its abode several times a year, or once a year, or every other year. The richest man in the Tarahumare [sic] country, now dead, had five caves, and moved as often as ten times in one year. (Lumholtz 1902:171)

Nearly all ethnographies on the Rarámuri describe their residential mobility. Although I was familiar with these ethnographies before I entered the field, I did not appreciate the importance and prevalence of various Rarámuri mobility strategies fully. In recording Rejogochi households' residential moves, it became clear that residential mobility is both high and quite varied (see also Merrill 1981).

Merrill (1981:50-55) records the following information for residential mobility in Rejogochi: in 1978, 28 households occupied Rejogochi for at least six months of the year. Of these 28 households, 24 changed residences at least once—18 moved to cold weather dwellings, and 13 moved during the agricultural cycle. Of the latter 13, 11 households moved to residences beyond Rejogochi, while two households moved within Rejogochi itself. Four other households moved to Rejogochi for between one and five months of 1978, three of these to tend agricultural fields. The fourth household had run out of food in its own community, and came to Rejogochi to work for subsistence for another Rarámuri household.

Table 1 lists all households who have residences in Rejogochi, who occupied them for any length of time during the tenure of my fieldwork in 1987 and 1988, and the general purpose of the occupation. Residential moves both within and beyond the valley are noted, although it is assumed not all cases are known. Thus, the data here are a conservative characterization of residential mobility by the inhabitants of Rejogochi. This is especially true for those households listed as not making any residential moves.

Based on Table 1, Table 2 summarizes the mobility strategies of households in Rejogochi. The 1987 estimate for residential moves per household (not including the moves to Basíhuare, the religious administrative center) is 1.5. Of the 51 residences I identified in Rejogochi in 1987, approximately 42 had habitable dwellings. Dividing this number by the total number of households with residences in Rejogochi, we could estimate about 1.4 residences per household. But Rejogochi is only one community within a larger settlement system that the Rarámuri throughout this region utilize. *Residences* in other communities, occupied by *residents* of Rejogochi have not been counted, while temporary inhabitants of the study community and their habitations have been.

A sampling of the range of household movements observed in Rejogochi demonstrates some of the ways the Rarámuri implement various land

Table 1. Household mobility strategies in Rejogochi.

Household Number	Main Residence in Rejogochi	Residential Mobility Biseasonal	Agricultural	Other	Notes
1	Y	Y	Y	Y	1
2	Y	Y	Y		
3	Y	Y	Y	Y	2
4	N	Y	Y		3
5	Y	Y			
6	Y	Y			
7	N	Y	Y		3
8	(N/A)				4
9	Y	Y			
10	Y	Y			5
11	N				6
12	Y				
13	Y	Y			7
14	Y				
15	Y				
16	Y	Y	Y	Y	8
17	Y	Y			
18	Y	Y			
19	Y	Y			
20	Y				
21	N	Y			3
22	Y				
23	Y				
24	Y	Y			
25	Y	Y			
26	N	Y	Y		9,10
27	Y				
28	Y				
29	Y				
30	Y				
31	N	Y	Y		3
32	N	Y			9
33	Y	Y			
34	N	Y			9

Notes

[1]Moved dwellings to new residence location
[2]Death of male household head
[3]Occupies residence periodically during agricultural cycle, or on occasional visits to relatives
[4]Household did not exist in 1987/88
[5]Only young married couple of household moves
[6]Household does not own land or residences in Rejogochi
[7]Established new household in 1987
[8]Other residential moves to work for mestizos
[9]Household does not own residence in Rejogochi
[10]Ran out of food in their main community

use and mobility strategies (see Table 3). The name given for each household member is a pseudonym. In the following discussion, Households 1, 3, 5, 6, and 10 were all households with whom I interacted frequently—up to several times a week. These were households I visited regularly, often participating in daily chores. Many members of these households, especially the women and children, were frequent guests at my home. I was less familiar with members of Household 2, although I visited them and attended *tesgüinadas* and fiestas at all of their residences in Rejogochi. They, too, occasionally visited me at my home. I had the least amount of interaction with Household 7 and Household 4. Household 7 was infrequently in Rejogochi to visit the female household head's parents and siblings (Household 6), and to tend her fields here. I interacted with this household only when our visits to the residences of Household 6 overlapped. Household 4 was also in Rejogochi only periodically. This household's main community was Bakéachi, a small community about two hours' walk from Rejogochi. I usually interacted with this household when its members were visiting or living with Household 3, but I visited it when I assisted in the harvesting of Household 3's fields, located near Household 4's residence, and during *tesgüinadas* it hosted in both Rejogochi and Bakéachi.

Household 8 lived in Rejogochi in 1983. My interaction was brief but intense while Robert J. Hard and I mapped its residence and noted activities and activity loci during a ten-day period in March 1983. Information from notes and comments on this household's mobility strategies is included here as an example of a year-round sedentary household (Graham, fieldnotes, 1983; R. Hard, personal communication, 1983, 1989; Hard and Merrill 1983; W. Merrill, personal communication, 1983, 1989).

Table 2. Summary data of household mobility strategies (see Table 1).

Total Number of Households in Rejogochi Between January 1987 and March 1988	32
Total Number of Households with Residences in Rejogochi	29
Number of Agricultural Residences	5
Households with Main Residence in Rejogochi	25
Households with Main Residence in Rejogochi Making Residential Moves	15
Total Number of Households Making Residential Moves	22
Households Making Biseasonal Moves	12
Households Making Agricultural Moves	16
Households Making Other Moves*	5
# Households/# Total Moves	1.5
# Households/# Biseasonal Moves	2.7
# Households/# Agricultural Moves	2.0
# Households/# Other Moves*	6.4
% Households Making:	
Residential Moves	68.7
Biseasonal Moves	37.5
Agricultural Moves	50.0
Other Moves*	15.6
% Main Residents Making Residential Moves	60

*Not including residential moves to religious center

Table 3. Households and residences in Rejogochi discussed in text

Houshold Number	Residence Number	Occupation Type	Other
1	1	'main'	occupied by entire household or by the household's older married couple and their unmarried (divorced) daughter
	2	warm weather	occupied mainly by son's family
	3	'main'	combined dwellings from Residences 1 and 2 to form this main residence for the entire household
2	4	cold weather	rockshelter
	5	warm weather	occupied by eldest son and family, and male and female household heads
	6	warm weather	occupied by second son and family, and occasionally by male and female household heads, and possibly their youngest children
	7	warm weather	for sleeping only, by male and female household heads
3	8	'main'	after death of household head in winter of 1987, residence was occupied only during the agricultural cycle
	9	cold weather	rockshelter used by various households (but not Household 3) during cold weather, and briefly during the agricultural cycle
4	10	agricultural cycle	household does not have a main residence in Rejogochi
5	11	'main'	(warm weather residence)
6	12	'main'	residence occupied during warm weather, and also used as a storage facility as occupation of Residence 14 increased
	13	cold weather	rockshelter
7	14	agricultural cycle/ 'main'	occupied by Household 7 periodically during the agricultural cycle, but also used during much of the year by Household 6
	15	cold weather	rockshelter housing winter storage structure of Household 6, sometimes occupied by Household 7 during cold weather
8	16	year-round	occupied in 1983; household heads are both dead, and the dwellings at this residence have been removed and established elsewhere
9	17	warm weather	
10	18	year-round	occupied in the past as a warm weather residence?
11	19	?	residence in Horírachi; household does not own land or residences in Rejogochi
12	20	year-round	household divided in 1987, with Household 14 moving to a new residence
13	21	residence	unoccupied in 1987 prior to its purchase, and its dwelling subsequently moved several months afterwards
	22	(new location)	dwelling from Residence 21 moved to this location in spring of 1988
14	23	year-round	household plants only in Rejogochi

Household 1—Ignacio Bitibachi and Margarita Upachi

Ignacio Bitibachi and Margarita Upachi live with their married son Patricio, his wife Chabela and two young children, and their unmarried (divorced) daughter, Mirita (Figure 11). Ten years ago this household wintered in Guaguéachi and lived in Rejogochi during much of the summer. It now occupies residences in Rejogochi almost full-time, going to Guaguéachi during the agricultural cycle, but spending only part of the winter there. During visits to Guaguéachi, this household's residences in Rejogochi may be abandoned for several days to two or three weeks at a time. At times, part of the household remain in Rejogochi while the rest go to Guaguéachi.

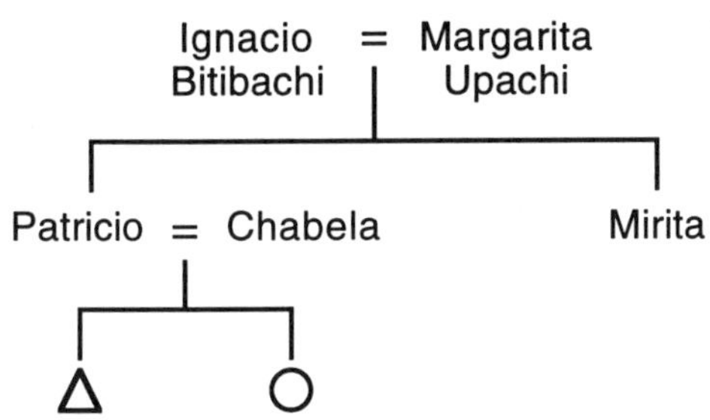

Figure 11. Household 1.

In the spring of 1987 this household had two residences in Rejogochi, separated by about 500 m and the perennial stream that runs through the valley (Residences 1 and 2, Table 3). Everyone used the house on the east side of the stream (Residence 1) at various times of the year, but Patricio and his family also lived in the second residence (Residence 2), located upstream on the opposite bank. In August 1987, Patricio moved the dwelling from Residence 2 to a new location about 300 m upslope and to the west (Structure A at Residence 3). In March of 1988, Patricio and Ignacio moved the dwelling from Residence 1 across the stream and put it next to the dwelling Patricio and Chabela used (Structure B at Residence 3). While Residences 1 and 2 were both located in close proximity to fields owned by members of this household, the new residence (Residence 3, Table 3) is at least 200 m from the household's nearest field. Of the two dwellings at this new residence, Structure A (from Residence 2) is used as private space for Patricio and his family, and Structure B (from Residence 1) is used by the older couple. Mirita, Ignacio's unmarried daughter, occupied both dwellings, but tended to sleep in the dwelling used by her parents. During Holy Week, 1988, Mirita married a widower, also living with his natal family in Rejogochi. This new couple's living arrangements were not established before I returned to the United States.

Patricio told me he moved the dwelling from Residence 2 because many of his children had died there, although the house had occupied its position for no more than five years (Graham and Hard 1983). Another man, responding to my question of why the household had moved the house, suggested it was in order for the household to have a better view of the valley. Residence 2 was situated in a low area surrounded by corn fields, and visibility was quite limited. Residence 3 is on a slope above the fields and has a good view of the valley.

Another factor that merits consideration here is the location of Residence 2 and the time at which it received morning sun during the winter. Residence 2 lies on the valley bottom near the perennial stream. On the opposite (east) side of the stream, the valley floor is relatively narrow. A mesa rises up abruptly from it, so that the sun must rise above the crest of the mountain before the dwelling and surrounding activity areas are warmed. During February and early March 1988, I mapped the abandoned Residence 2. In the mornings, the frost did not melt until the sun crested the mesa and warmed this location, between 9 and 10 a.m. Residence 3 received sun 60 to 90 minutes earlier.

In 1987, Residence 2 was *not* occupied during the winter; Residence 1 and possibly dwellings in Guaguéachi were used, instead. Nevertheless, by moving the structure from Residence 2 (and also from Residence 1) farther west and upslope, the new residence became more suitable for habitation for a greater portion of the year.

In 1987 this household cared for the goat herd of a local mestizo. Whenever Household 1 was in Rejogochi, the animals were sheltered in a recently abandoned house. (This barn is counted as one of the uninhabited residences in Rejogochi, and not numbered here.) The barn provided the herd protection from the weather and predators. It was close enough to the household's various residences that a move to a separate cold weather residence for the sake of the animals (for example, the household's rockshelter in Guaguéachi) was not necessary.

Thus, in the past Household 1 had made more moves between their various cold weather residences, as well as moving during the agricultural cycle, than currently. But the shift in emphasis of the main residence from Guaguéachi to Rejogochi, combined with adequate local accommodations for the herd animals during the winter has resulted in this household establishing a residence more suitable for year-round occupation. Whether Household 1 will continue to use Residence 3 in this way, of course, remains to be seen. The possibility of changing residence locations for other reasons, e.g., infestation by rodents and vermin, are here recognized; however, these were not discussed with the household.

In both 1987 and 1988 Patricio was appointed a *Phariseo* (Pharisee). Pharisees are one of the two groups that arrange the Holy Week ceremonies; see Kennedy and Lopez (1981) and Merrill (n.d.) for detailed discussions of this role. This office entails a number of duties, and it is possible that Household 1 elected to spend the winter in Rejogochi, because it is about an hour closer to Basíhuare than is Guaguéachi. Unlike households of other religious officials in 1988, Household 1 waited until relatively late before making the residential move to Basíhuare for Semana Santa. I suspect this delay was in part due to their responsibilities for the large herd of goats; the general care-taking of these animals, and finding pasture for them, is time consuming. In addition, there is probably more pasturage around Rejogochi than near Basíhuare at this time of year, when new grasses have not yet sprouted in abundance, and a larger than usual number of herds are grazing in and around the latter. Members of some households with large sheep and goat herds say they do not participate in events at the religious-administrative center because of these logistical problems.

Household 2—Valencio Rosákachi and Tenchi Chuwina

Valencio Rosákachi and Tenchi Chuwina have one of the largest extended families in Rejogochi (Figure 12). In addition to four unmarried children, they have two married sons and a married daughter, all of whom make up part of the co-residential group of Household 2 for at least part of the year. Juan Sébio, the oldest son, is married to Virginia, and they have two young daughters. Jacinto, the next oldest son, and Lucrécia, his wife, have a baby daughter. Jesusita is married to Maurício and they also have a young daughter.

During the winter, Valencio, Tenchi, and their unmarried children occupy a rockshelter with Juan Sébio, Virginia, and their children (Residence 4, Table 3). They are joined for part of the winter by Jesusita and her family. During warm weather, Juan Sébio and Virginia occupy a house in the lower valley as their primary residence (Residence 5, Table 3), while Jacinto and Lucrécia have a warm weather dwelling in the upper valley (Residence 6, Table 3). Valencio and Tenchi, with the younger children, divide their time between these two residences during warm weather. The married daughter, Jesusita, and her family lived in Residences 4 and 5, and also with Maurício's parents at their residence.

The warm weather residences (Residences 5 and 6) are both associated with fields owned by the household. Valencio owns both residences, and has given or lent them to his sons. Juan Sébio is building a new house (in construction for over a year)

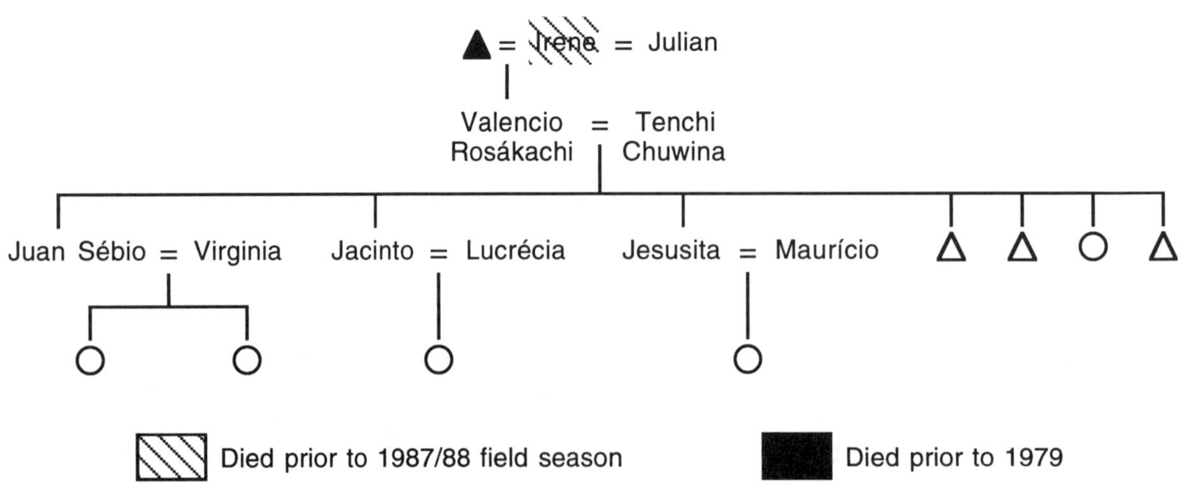

Figure 12. Household 2.

about 350 m east of Residence 5, across the perennial stream from it. He says when that dwelling is finished, Residence 5 will again be occupied by his father.

Currently, Valencio and Tenchi sleep in a dwelling (Residence 7, Table 3) located about 150-200 m downslope of Residence 5 during warm weather, but their "living" space is at Residence 5 and no household goods other than sleeping boards are usually found at the structure at Residence 7. In the winter of 1987/88, much of the structure was disassembled and used to create additional sheltered space at the household's rockshelter (Residence 4). Towards the end of the winter, the dwelling from Residence 7 was reconstructed in its original location. While the Rarámuri will move structures from one place to another, as described above for Household 1, usually they do not disassemble and move them *seasonally*. The structure at Residence 7 was little more than a lean-to, and its boards provided the structural elements to create additional sheltered workspace at the rockshelter (Residence 4).

Valencio works land only in Rejogochi; other fields he had owned and planted outside of this community he has given to Juan Sébio and Jacinto. Juan Sébio said both he and his brother cultivate these more distant fields. At Juan Sébio's holdings, there is a house and storage structure used when tending that land. I do not know how many trips he made there, or the length of these visits. In 1988, Juan Sébio and Virginia also spent a couple of weeks (or more) in Virginia's natal community, helping her parents to plow and co-hosting a drinking party. Possibly, Virginia owns land there, also cultivated by the nuclear family.

Household 2 owns a number of herd animals which the various household members, both adults and children, take turns watching. The sheep and goats stay in a small corral at the rockshelter in the winter.

Household 2, except for Jesusita and her husband and child, shared a single rockshelter during Holy Week in 1987 and 1988, and also during the Fiesta de Guadalupe in 1987 in Basíhuare. Jesusita and Maurício shared a different rockshelter with Maurício's natal household and a third household for the 1987 Holy week. During the 1988 festivities they stayed with the rest of Household 2 and a third household, in "their" rockshelter in Basíhuare.

As household structure changes (e.g., as households go through the domestic cycle and the number of individuals and available labor force increases or decreases), a number of changes are also possible in housing accommodations. That is, as household structure changes, so do aspects of residential site structure.

The household of Valencio and Tenchi (Household 2) is a case in point. As children of this household have married, its members use several residences. For example, in 1977/78, Residence 5 (Figure 13) was occupied as the main residence for Household 2. In 1978, Residence 6 (Table 3) was used for increasing periods of time, until by the winter of 1978/79 Residence 6 was the main residence for the household, with Juan Sébio sometimes occupying Residence 5 after his marriage. Another household, moving temporarily to Rejogochi in 1978/79, also occupied Residence 5 for a short time (Hard and Merrill 1983).

During 1987/88, Household 2 continued to divide and coalesce. The brothers Juan Sébio and Jacinto occupied residences at different ends of the valley for most of the year, and Valencio and Tenchi moved between these residences (Residences 5 and 6). Moreover, Valencio and Tenchi (and probably their two youngest children), occupied another structure as a sleeping facility (Residence 7) while performing all other activities at Residence 5.

As stated above, Juan Sébio, the eldest son of Household 2, is currently building his own house about 350 m from Residence 5 where he will live with his wife and children. Valencio, Tenchi, and their unmarried children will then occupy Residence 5 as both sleeping and living space. The new dwelling is being built on the site of a residence occupied by Juan Sébio's grandmother Irene, and her husband Julian, in the late 1970s (Hard and Merrill 1983).

Young households do not always build their own residences. In January 1987, Household 12 consisted of Lirio Rosákachi and Inez Bitibachi, their unmarried child, and Lirio's married daughter Marisanta, her husband Lupe, and Marisanta's three young children (Figure 14). Household 12 lived year-round at Residence 20 (Table 3). In spring 1987, Lupe bought a dwelling belonging to another household, that did not use that residence (Residence 21, Table 3). Marisanta and her husband and children became a separate household when they moved into this residence (Household 13, Figure 14).

In February 1988, Lupe and Marisanta hosted a drinking party to move the dwelling from Residence 21 across the stream to its current location (Residence 22, Table 3). This location receives sun in the winter several hours earlier than did the original residence (Residence 21). It is also closer to some of the household's fields, and within 150 m of Residence 20, where Lirio and Inez (Household 12) live.

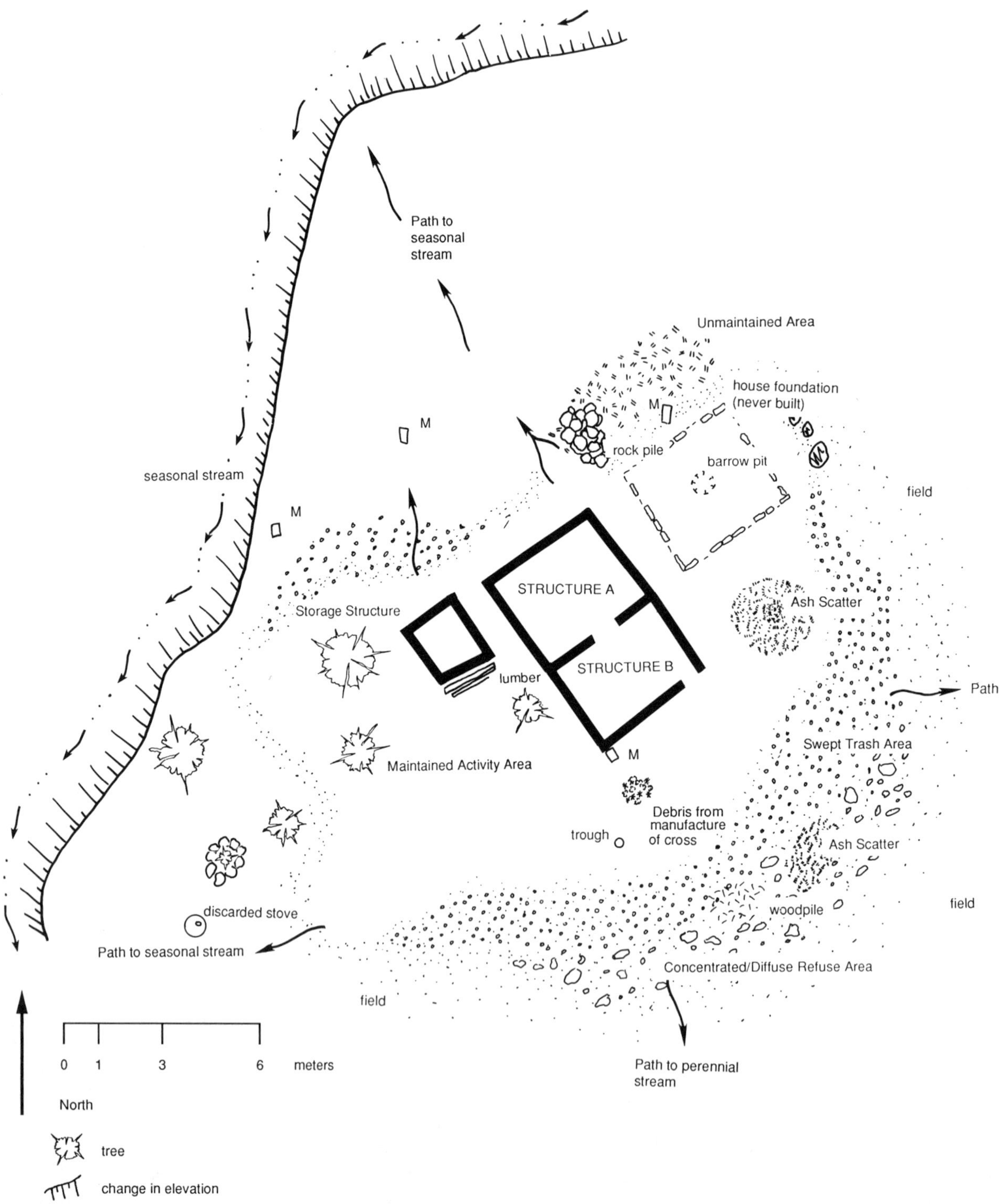

Figure 13. Residence 5.

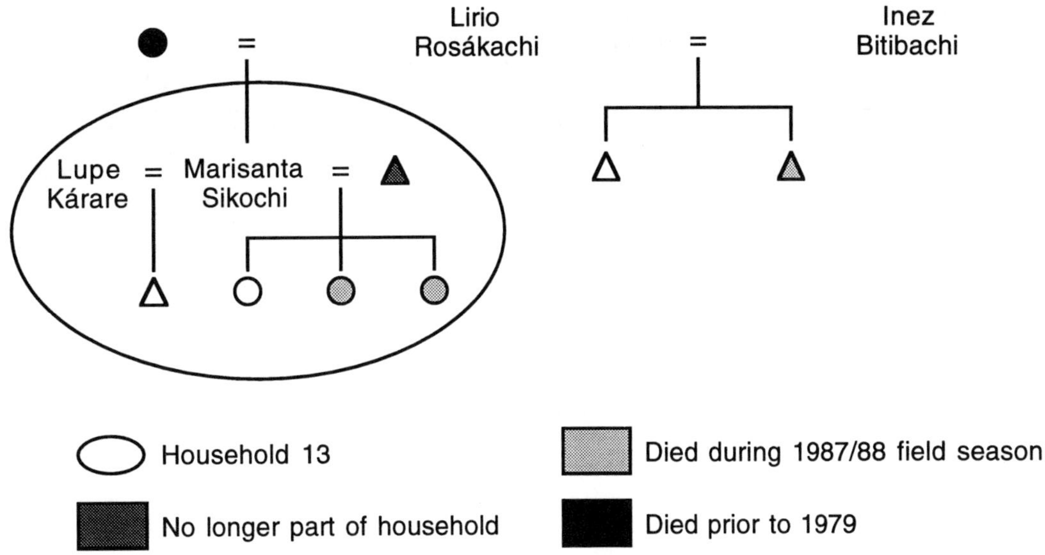

Figure 14. Households 12 and 13.

Household 3—Rafael Kapírachi and Santa Rosa Sitánachi; and Household 4—Alejandro Watagéachi and Graciana Rikalachi

Rafael Kapírachi and Santa Rosa Sitánachi and their five young children made up Household 3 in 1987 (Figure 15). This household occupied a residence in Rejogochi (Residence 8, Table 3) for most of the year, moving to a residence in the community of Ariséachi for a couple of weeks in the fall to harvest crops planted there. For part of the winter of 1987/1988, the household shared a rockshelter in Bakéachi with Santa Rosa's natal household.

Rafael was quite ill, and died in Bakéachi during the winter of 1987. Because of his illness, the household enlisted the help of relatives to help plow and plant in both Rejogochi and Ariséachi. The household postponed several trips to Ariséachi during the growing season because of Rafael's illness; instead, Santa Rosa or her father would make day trips to tend this field. At harvest time, however, the entire household did move to Ariséachi for a couple of weeks, occupying the house Rafael had inherited there. Upon Rafael's death, the surviving members of the household returned to Rejogochi, staying with relatives, first in their rockshelter and later in their main residence. Santa Rosa and her children then reoccupied Household 3's main residence (Residence 8) with Household 4 (see below) when the latter came to Rejogochi to plow their fields (both those of Household 3 and 4). Santa Rosa then took her children and returned to Bakéachi with Household 4, and was residing there with her natal household at the time I returned to the United States.

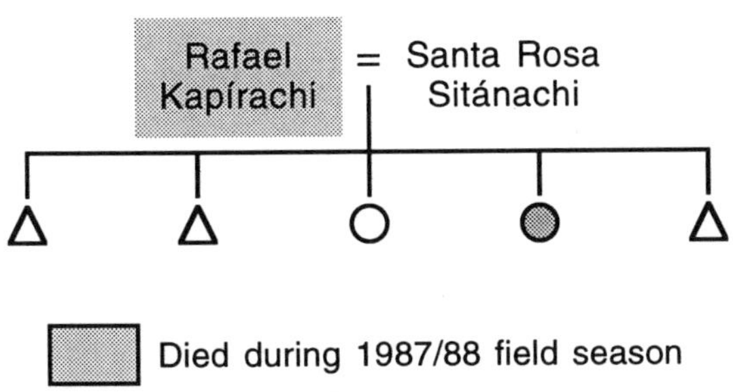

Figure 15. Household 3.

Household 4 is an extended family headed by Alejandro Watagéachi and Graciana Rikalachi, with their married son, Santoreyo, his wife Carolina, and their children, and their married daughter, Chepa, her husband Antonio, and their small children (Figure 16). This household stays in Rejogochi infrequently; its main residences are in the community of Bakéachi. When Alejandro and Graciana move to Rejogochi, they live near one of their fields (Residence 10, Table 3). At these times, some of the household (e.g., Santoreyo and some of

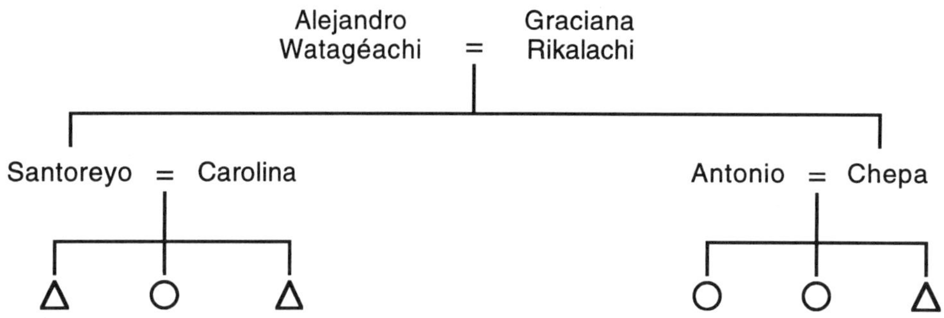

Figure 16. Household 4.

his family) might also stay with Santa Rosa and Rafael at their home (Household 3, Residence 8), with whom Household 4 is related.

Antonio was born in Rejogochi, and he and his wife and children occasionally stay at his parents' residence, including times when other members of Household 4 are not in Rejogochi. When I left Rejogochi in April 1988, Antonio and Chepa were beginning to construct their own dwelling at his parents' residence.

After Rafael's death (Household 3), Household 4 occupied both Residence 8 and Residence 9, a rockshelter belonging to Rafael. This rockshelter was also occupied briefly by Household 5 (see below) during the winter of 1987/1988. That household may have abandoned Residence 9 so that Alejandro and Graciana and their family could occupy the shelter, although this was never explicitly stated. The rockshelter had partially collapsed early in 1983 while another household was occupying it. Although several people were seriously injured, no one died in the rock fall. Since then, Residence 9 has been occupied only infrequently (and cautiously).

Santa Rosa's parents from Bakéachi also own and cultivate land in Rejogochi. Unlike many households, this couple generally do not make extended stays in Rejogochi when tending their fields probably in part because they do not have a structure built there. Instead, they make day trips, or occupy the rockshelter (Residence 9) if they stay the night. Thus this rockshelter is used by a variety of households both as a cold weather residence and as an overnight camp during the agricultural cycle. As the discussions of Residences 3 and 9 show, ownership of rockshelters is not readily apparent.

Rafael was too ill to go to Basíhuare for Holy week in 1987, and the household remained in Rejogochi. In 1988, Santa Rosa and her children stayed with relatives in Bakéachi and Ariséachi (both of which are closer to Basíhuare than is Rejogochi), participating to a limited extent in the ceremonies and fiestas held in Basíhuare.

Household 3 demonstrates how illness or incapacitation can be important factors in determining where a household will spend all or part of any given year. Rafael and Santa Rosa did not spend the winter of 1986/1987 in Bakéachi, for example, because Rafael was too ill to make the trip. They were afraid to use Residence 9, their own rockshelter in Rejogochi, because of the 1983 collapse. Other households have also not made residential moves during the agricultural cycle under similar circumstances of serious illness or debilitation within the household. The relative emphasis of "primary" and "secondary" communities may also change as a result of these factors.

The death of one or more family members also may result in significant changes at a household's residence, as well as their mobility strategies. Sometimes a residence is abandoned. At the most extreme, a household will completely dismantle and sell or give the dwelling to another household which then establishes a residence in a new location (this is described below for Residence 16 of Household 8). In other cases, if a household member frequently dreams of a dead family member (who died at that residence), the dwelling may be moved to another location, the household establishing a new residence using the structural elements from the older one. Patricio (Household 1) gave the death of many children as the reason for relocating the dwelling from Residence 2 to Residence 3 (Structure A).

Household 5—Sorpino Rewesuchi and María Elena Batiana

Sorpino Rewesuchi and María Elena Batiana live year-round in Rejogochi with their five unmarried children (Figure 17). Household 5 plants only within the valley, although it may own land in another community. It occupies its main residence on the valley floor for most of the year (Residence 11, Table 3), but for at least part of most winters it moves to a rockshelter. During the winters of 1986/1987 and 1987/1988 the household lived in Residence 9, the rockshelter often used by Household 3. At that time, María Elena cared for a relative's flock of sheep and she needed to provide them with warm shelter during the cold winter nights.

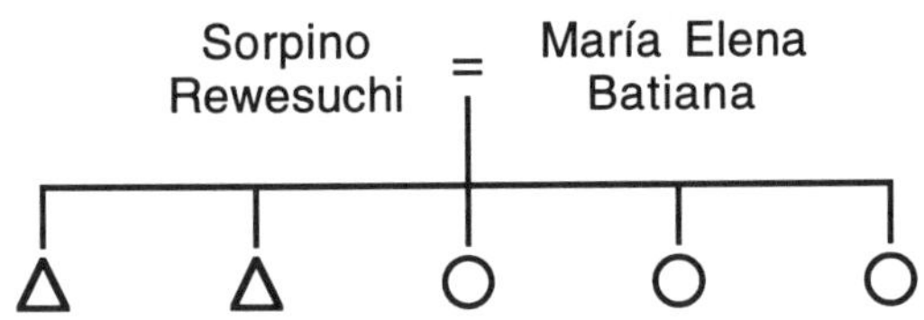

Figure 17. Household 5.

The path between the rockshelter and the household's main residence led past my own dwelling in Rejogochi. During the winter I observed almost daily trips, often more than one trip a day, between the two residences to pick up some piece of equipment, to retrieve more food supplies, or to carry out some activity at the main residence.

Sorpino was often away from the community, engaged in various forms of wage labor. On occasion during his absences, the rest of the household would spend a night or two with either Sorpino's sister's household or a relative of María Elena's, both of whom live in Rejogochi. These overnight stays at relatives' residences within Rejogochi are not considered examples of residential mobility in this study. No moves were made by this household during the agricultural cycle.

Household 6—Calistro Rosákachi and Guadalupe Sitánachi; and Household 7—José María Raramuchi and Sarina Basachi

Calistro Rosákachi and Guadalupe Sitánachi, with their four children (Household 6) live year-round in Rejogochi and are joined periodically by their eldest daughter, Sarina Basachi, her husband, José María Raramuchi, and their young children (Household 7, Figure 18). Calistro built Residence 12 (Figure 19) 20 to 25 years ago. About 1986, Sarina and José María (Household 7) built Residence 14, a log house ca. 75 m up valley (north) of Residence 12. In the current use patterns of the residences, Household 6 uses Sarina's home at least as much as, if not more than, its own dwelling, regardless of whether or not Sarina is in residence.

During cold weather Calistro and Guadalupe move their family to a rockshelter (Residence 13, Table 3), as they have every winter for about 25 years. It is interesting to note that whereas the previously discussed rockshelters, Residence 4 (of Household 2) and Residence 9 (of Household 3), have had a number of different households as residents during the past ten years, the sole occupants of Residence 13 have been the members of House-

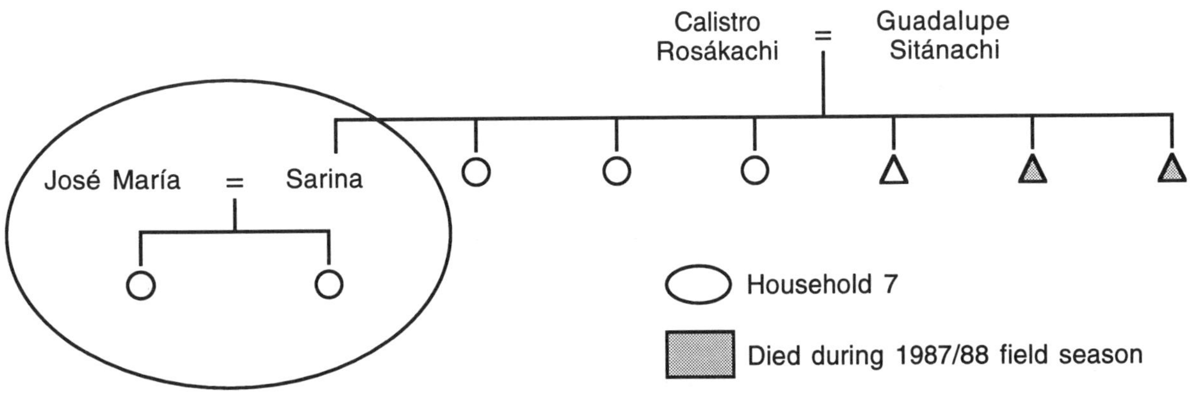

Figure 18. Households 6 and 7.

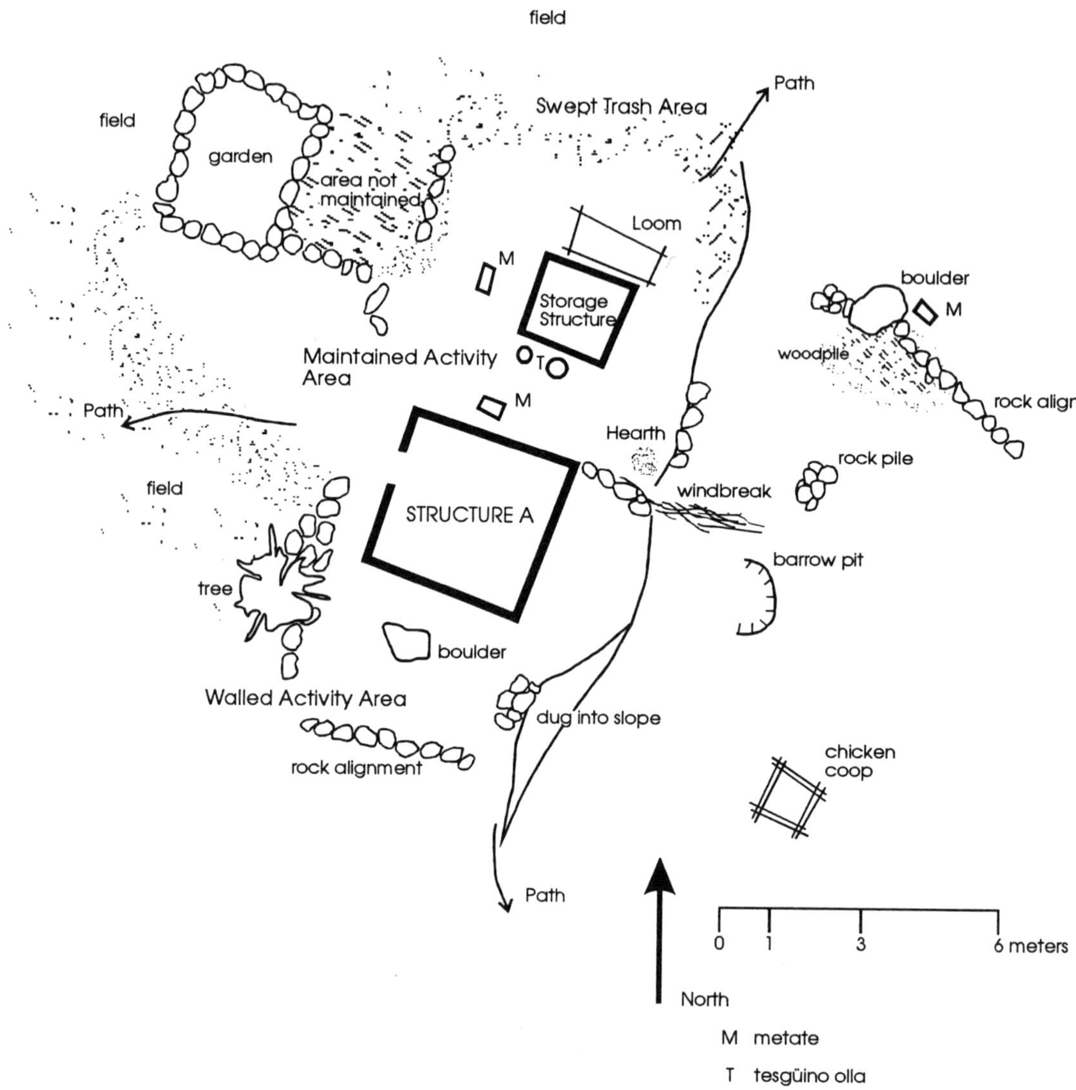

Figure 19. Residence 12.

hold 6 (Hard and Merrill 1983). Whereas the other rockshelters discussed here have been used for two or more generations, Calistro probably was the first to modify the area at Residence 13 to make it a suitable winter residence (W. Merrill, personal communication, 1992).

In general, more goods are stored year-round at this rockshelter than at either Residences 2 or 9. For example, Guadalupe made a second loom in 1987, and now has one at the main residence and the other at the rockshelter. Wool, baskets, cookware, and other items are stored in the storage structure, or cached in and around the structures at Residence 13. The storage structure is about 25 to 30 m away from the habitation, in its own, smaller, rockshelter. This shelter is infrequently used as residential space as well, most often by Sarina and her family (Residence 15, Table 3).

Currently Household 7's main residence is in José María's natal community, but the household makes extended residential moves to Rejogochi throughout the year. Even when the family is not in residence in Rejogochi, Household 6 now spends the majority of its time living at Residence 14. I was unable to get a satisfactory answer for why the latter residence is preferred, but I suspect it is due in part simply to the older age and beginnings of decay of the structural elements of the dwelling. Insect and vermin infestation at Residence 12 may also be factors in its more limited occupation.

During Holy Week, 1987, Calistro and Guadalupe, and José María and Sarina lived in an abandoned adobe house for more than a week prior to Easter Sunday. For the 1988 ceremonies, the two households occupied a room in the abandoned mule station in Basíhuare. José María was one of the Pharisees for the 1988 Easter ceremonies, and Household 7 lived in Basíhuare for longer than did Household 6.

As a household matures, and children marry and establish their own dwellings, a number of factors come into play regarding the placement of that new dwelling. Conflict between generations, or the lack thereof, is an important consideration. Crowding, and physical constraints on available space is also important.

Residences 12 and 14 are both built between the edges of fields with the base of the slope from the mountain behind. There was no room to build Residence 14 nearer Residence 12 without turning part of the field into the generalized activity area. Residence 14 stands in an area that has apparently never been utilized as cultivated land (Graham and Hard 1983).

This notion of not taking up agricultural land with residential compounds is reinforced by Household 1's (Table 3) placement of residences. Neither Residence 1 nor Residence 2 were located in areas with sufficient available land to allow for the building of a second dwelling at the residence. Residence 3 is located in an area where there are no fields. Although Residence 3 joins two separate residences into one, the two units of Household 1 shared residential space to the extent that they were in fact considered a single household, rather than two, even before they combined the two dwellings at a single residence.

Household 8—Cruz Akichi and Irene Kárare

In 1983, Cruz Akichi and Irene Kárare (Household 8, Figure 20) lived with their young son in Residence 16 (Figure 21, Table 3). Prior to about 1977, Household 8 had lived with Cruz's natal

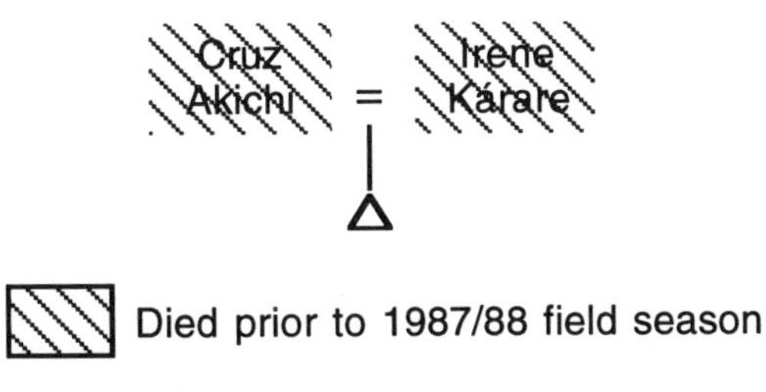

Figure 20. Household 8.

household in a residence approximately 200 m upstream from Residence 16. Residence 16 at that time was occupied by Cruz's maternal grandmother, and consisted of a single dwelling (Figure 21, Structure A). When Cruz's grandmother moved over to his natal residence, presumably because she was too old and feeble to live by herself, Cruz and Irene moved into the house, and added a second structure (Figure 21, Structure B).

Household 8 combined its herd with those of Cruz's natal household. The animals occupied an abandoned house as a barn, and the two households shared the care-taking responsibilities of the animals. The presence of the barn and the relatively large consolidated land holdings of Household 8 are two factors that allowed Cruz and Irene to remain year-round at Residence 16.

In late 1983 or early 1984, Cruz died. Irene remarried another member of the community, and together with her son moved away from Residence 16. Within the next year, Irene also died, and the surviving member of Household 8, the child, now lives with his paternal grandparents.

By 1987, Residence 16 no longer existed. Between 1984 and 1987, Cruz's parents changed their residence from the home they had occupied from before 1977, and their youngest son, Cruz's younger brother moved the two structures from Residence 16 to a site farther downstream, close to the new residence. In the late winter/early spring of 1987, the two structures from Residence 16 both faced outward, whereas when occupied by Household 8, Structure B opened into Structure A. During 1988, one structure was moved upslope, away from the residence, and now probably serves as a storage unit, while the other structure remains as a dwelling for Cruz's brother's household.

Household 10—Mariano Okowíchi and Juanita Chiwáware

Household 10 consists of Mariano Okowíchi and Juanita Chiwáware, their married daughter and her husband (Keresita and Candelario) and their two children, Mariano and Juanita's four unmarried children, and also Mariano's mother Katilachi and her husband Cornelio (Figure 22). Household 10 has occupied their residence (Residence 18, Table 3 and Figure 23) for a number of years. During previous winters the household sometimes lived in various rockshelters in the area. It is not clear whether the whole household always remained together at these times, or if some of the members remained at Residence 18 (Hard and

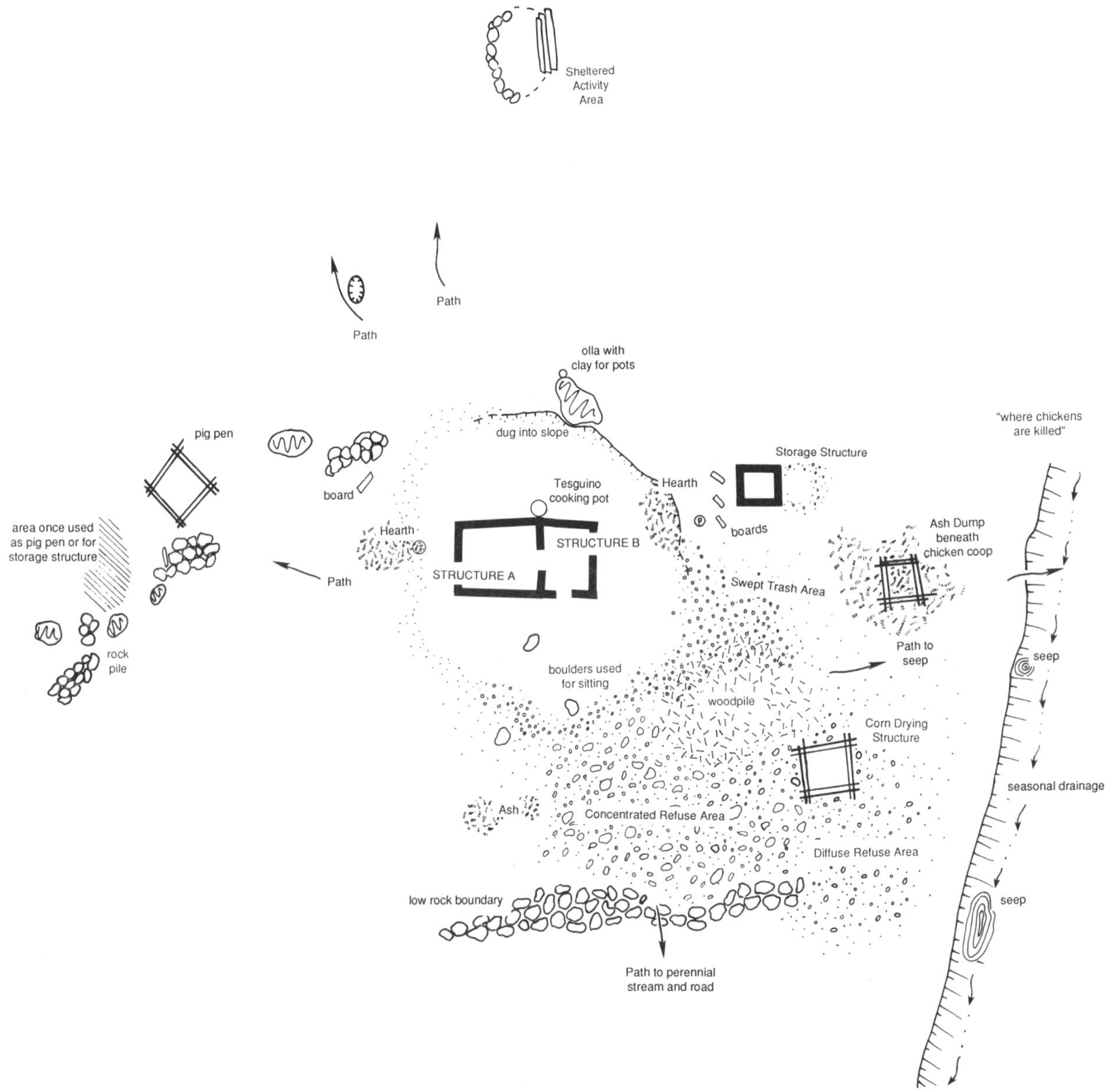

Figure 21. Residence 16 (from Robert J. Hard, with permission).

Merrill 1983). In January 1983, Household 10 was living in Residence 9 when a portion of the rockshelter collapsed, injuring Katilachi, Cornelio, and two of the young children. To my knowledge, Household 10 has not made a biseasonal residential move since that time.

Until several years ago, Structures B and C were the only dwellings at Residence 18 (Figure 23). Structure B was used as the main living space, with Structure C used for storage and occasionally for sleeping, most frequently by the children of the household (Hard and Merrill 1983). The walled activity area was used for cooking and also sleeping. The area identified as a goat pen in 1987/88 was used as outdoor sleeping space in 1977-1979 (Hard and Merrill 1983).

Keresita married in 1983, and she and Candelario continue to move back and forth between their natal residences. Probably in early 1984, the young couple began constructing a dwelling in Rejogochi at Residence 18. They were still modifying the structure, Structure A, when I began my fieldwork in January 1987. During that year, the walls of the house were raised higher, the roof completed, and a solid door constructed. Prior to its completion, the structure had been in use for some time. Household 10 continued to modify Residence 18 as a whole as a result of this new structure and the addition of the stored goods and household assemblage of the newest nuclear family.

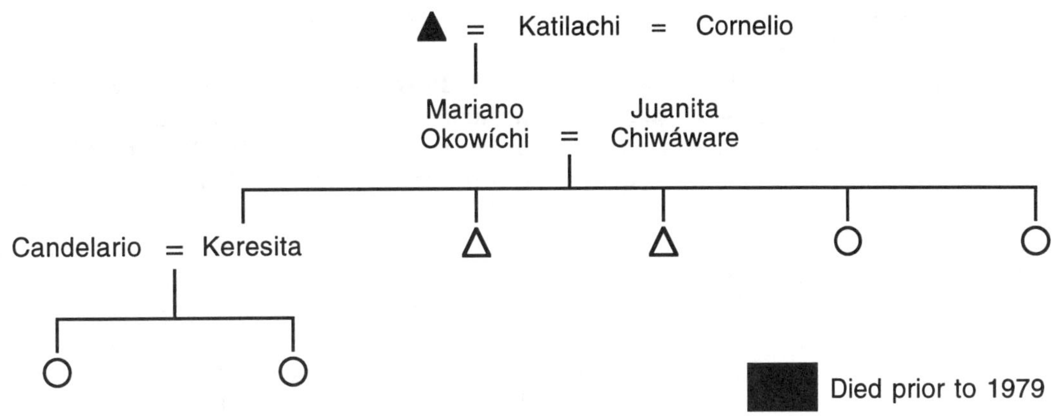

Figure 22. Household 10.

Figure 23. Residence 18.

Cornelio added and later dismantled a second storage structure in the northwesternmost area of the residence, and the storage structure shown in Figure 23 was moved farther west. The household also constructed a corn drying structure north of the fence along its eastern extension. A second outdoor cooking and sleeping space, already discussed, was created west of Structure A.

Although Residence 18 appears to belong to an integrated household, in reality at least two semi-discrete economic units live there. Structure B and the walled activity area are the domain of Katilachi and Cornelio, whereas Structure A contains the goods and activities of Keresita and Candelario when they are in residence, and Mariano and Juanita at all times of the year. Katilachi is an important care-taker of her youngest grandchildren and great-grandchildren, and the young children of the household move freely between the dwellings and their residents.

Unlike Households 1, or 6 and 7, when Household 10 added Structure A (Figure 23), curtailing any amount of agricultural land was not an issue. The area where Structure A is now was probably part of the maintained activity area, or possibly of the general trash area, rather than the field.

Household 14—San Pedro Rosákachi and Marcelina Bakéachi

In 1987, Household 14 consisted of San Pedro Rosákachi and Marcelina Bakéachi, and their six unmarried children (Figure 24). In 1977, however, the household was much larger, including San Pedro's young sister, Chepa, and their maternal grandparents, Felipe and Marta. In 1977, this household lived on a low ridge above the valley floor in Rejogochi. Its dwelling was the old stone house (now a three-sided and unroofed sheltered activity area) at the location of its present residence (Residence 20). The household constructed a new house at the base of the ridge in 1978, covering it with the roof of the stone house. San Pedro said that they moved because he was dreaming of his dead mother (Hard and Merrill 1983). Another man told me it was because it was too cold on top of the low ridge. From 1978 to 1981, additional dwellings were added to the residence at the base of the ridge (Hard and Merrill 1983). By 1987, all but the oldest of these structures had been moved back up on top of the ridge. That structure remains at its 1978 location, used as a barn.

In sum, residences go through a number of

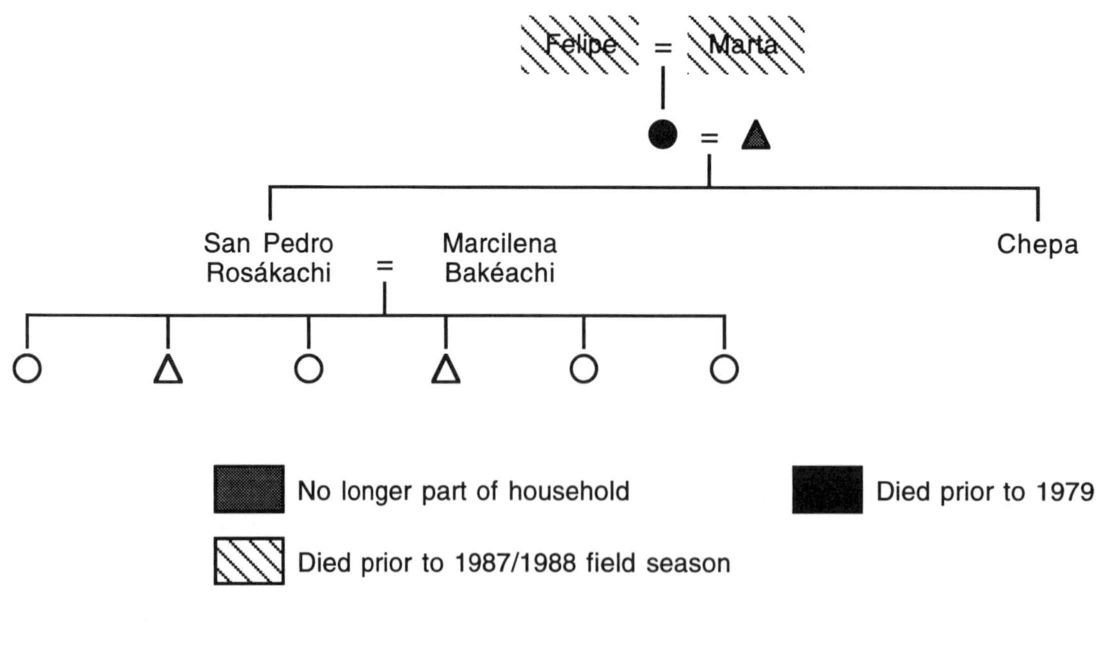

Figure 24. Household 14.

changes throughout their life histories in response to changes in the development of the household. The nature of dwellings and other considerations also influence how residences are used, how they are modified, and when they are abandoned. Rarámuri in Rejogochi consider houses constructed 25-30 years ago as quite old, and prefer newer dwellings. The changes in residential space as a result of changes in household structure occur in the context of two processes: accretional, as new structures and activity areas are added to existing residences, and subtractive, as household size decreases and structures are moved away from residence locations, or as the residence is abandoned completely. Different spatial requirements and constraints apply to *groups* of people than operate for *individuals*. The amount of space needed by an individual to perform an activity changes when that individual is working in the context of a number of other people. The same amount of space must be used differently as scheduling pressures change with changes in the number of workers in that space.

The Rarámuri Household and Labor Organization

One question I raised prior to my fieldwork was whether households join together at particular points during the year if an individual household is too small to accomplish a task—e.g., plowing or harvesting. Generally this is not the case, but a household may include a friend or relative for a short period of time for assistance. Girls in the last years before they marry assist women of other households. These young women will take over some of the childcare responsibilities of mothers with many small children and assist with basic subsistence activities. While neither strong enough nor skilled enough to be considered primary producers, these young women are effective at lessening the activity load of their mothers, relatives, and friends. Making corn beer and preparing food for work parties are other examples where a woman may call on the assistance of friends or relatives. While the examples given refer to women, the same cooperation in labor applies among Rarámuri men.

In very unusual circumstances, an individual in the community can be left with virtually no relatives to aid him in subsistence activities. This was the case several years ago with Julian, the husband of Irene (W. Merrill, personal communication, 1989; Hard and Merrill 1983). Irene was Valencio's mother (Household 2, Figure 12). For as long as Irene was alive, she and Julian lived with Household 2. Rejogochi was not Julian's natal community, and he had no close relatives in the valley beyond those acquired through marriage. Julian remained in Rejogochi for some time after Irene's death in the house in which they had lived (on the site where Juan Sébio is now building a new dwelling). During this time his interaction with Household 2 was minimal, and he shared meals and participated in most other residential activities with a different, neighboring household. By 1987, Julian had moved away from Rejogochi permanently, to live with his sister in another community.

When help for various activities is required, assistance is solicited from within one's network of relatives and friends. The co-operative labor groups described for Rejogochi quickly take over in performing many tasks. One can perhaps speak of a hierarchy of relationships within these groups. While households assist each other with particular tasks during the day, they rarely share the hosts' residential space. Rather, they each maintain their own residence. If they become a single co-residential group, it is generally only for a matter of days or, during religious festivities in Basíhuare, for two or three weeks.

Under one important set of circumstances a too-small household will join with another to form a larger, single household. This occurs at the death (or divorce or abandonment) of a spouse. In such an event, the surviving members of the nuclear family, if living as a distinct household, will join relatives and participate in many of the social and economic activities of that household (see discussion of Household 3 above). Depending on the circumstances, the bereaved may move several times, staying with different relatives and asking these various households for assistance in tasks and activities normally performed or organized by the deceased. When the single spouse remarries, these co-residential formations dissolve.

Chapter 3

Rarámuri Settlement and Residential Site Structure

Introduction

Individual Rarámuri household mobility strategies are complex and varied. The motive for presenting case examples here is to emphasize that, although we can assume a high degree of variation in mobility strategies, we must identify patterning at the level of the system rather than of the individual household. The ethnoarchaeological literature abounds with examples of singular strategies that I think would be impossible to recognize archaeologically. The questions that must now be asked are— what are the material consequences of Rarámuri settlement organization; what are the common assumptions archaeologists are likely to make, and how can they be better evaluated? Variations in archaeological "samples" of a cultural system are due to individual household size or differences in site occupation histories, as well as the technological and economic differences (O'Connell 1987; Oswald 1987; Rodman 1985; Wilk 1983, 1984).

Overriding individual household variation, however, is a broader pattern relating to the general cultural system. Here, I use observations of individual households' mobility strategies to extrapolate to a general pattern. In contrast with the specific case examples given above, this chapter concentrates on broad-based patterning when discussing archaeological implications and expectations of the Rarámuri settlement system.

The Regional Settlement Pattern

Typically, Rarámuri communities comprise a number of scattered houses, although the size and shape of residential complexes, the number of houses, and the distances between them vary considerably (for example, Figures 3 and 10, and Kennedy 1978:Figures 1 and 2, pp 36 and 40). In the barrancas, warm weather communities are commonly on the tops and sides of mountains, and rockshelters and other cold weather dwellings are down off the mountain tops. In the transitional zone where Rejogochi is located, houses are on or near valley bottoms, and rockshelters and other cold weather dwellings usually are on slopes or knolls above the valley floor. In the less extreme topography in the uplands, valleys are more open, and rockshelters are at the contact point between the valley floor and the low surrounding hills. The distribution of Rarámuri communities, and their size, therefore depends in part on local topography.

While the settlement *pattern* appears distinct between the three regions, in fact the organization remains the same: main and agricultural residences are typically located near fields. For the uplands and transitional zones, arable land is in the valley bottoms and on low knolls. In the barrancas, many fields are on mountain tops because the canyons are so steep and narrow that very little cultivation is possible in the valley bottoms themselves. The locations of both rockshelters and good water sources are critical factors in determining where people will spend their winters. Cold weather residences other than rockshelters are also located on knolls and low rises above the valley bottoms. These residences are chosen for getting good, early morning sun. They are generally south- to south-east-facing, and not overshadowed by the opposing mountain side. In the transitional zone and the barrancas, locating the dwelling above the valley bottom and below the mountain tops minimizes wind exposure and raises the residence above the layer of cold air that settles on the valley floor.

Residential Site Types

As well as the distinction between warm and cold weather dwellings (for example, Figure 9), many Rarámuri communities have some residences occupied only during the agricultural cycle, others occupied for the greater part of the year, and still others occupied year-round (Figure 25). The characteristic assemblages of the three basic types of residences—the main residence, the residence occupied during the agricultural cycle, and the cold weather residence—are discussed below. The residences described represent a certain type of occupation, but at any given time their function may change—for example, from a main to a part-time residence. From an archaeological perspective, this fluidity in site function hampers our ability to distinguish between the different types of residential sites. The importance of this difficulty will be explored in the discussion below.

Main Residences

The main residence is occupied for most or all of the year, and serves as the home base for a household (Residence 5, Figure 13). A garden is planted in the refuse area surrounding it. Storage structures hold most of the years' dried food stores. Figure 26 illustrates the interiors of the structures at Residence 5 (Table 3, and discussion of Household 2 above). Structure B of Residence 5 is roofed, with walls to a height of approximately 1.30 m. The walls of this structure are logs cut and processed by the Rarámuri. There is little chinking between the logs, and because of the open spaces above the walls and between the logs, the interior of the structure is well lighted. Structure A of Residence 5 is a well-chinked dwelling, also made of hand-hewn logs. Visibility within this structure is consider-

Figure 25. Rarámuri main residence. The two structures in front of the house were animal pens used to house pigs and turkeys, then the smaller structure (at right) protected a fruit tree sapling and some garden vegetables from grazing animals. This smaller pen is in the Concentrated Refuse Area, which ends at the line of rocks in the foreground of the photograph. The rock alignment serves as water control, preventing erosion of the small drainage to the right of the area in the photograph. It also limits the spread of trash from the residence. The household plants corn, beans, and squash in the part of the Concentrated Refuse Area shown in this photograph. The Maintained Activity Area is barely visible above the Concentrated Refuse Area; it can be distinguished only by the sharp contrast of its clean, bare white surface to the darker area below it. (See Graham 1993:Figure 3.2, p. 27, for a view of this residence during the agricultural cycle).

ably lower than in Structure B, especially on the rare occasions when the door is shut against inclement weather. Structure A has as its source of heat and for cooking a stove made of a 55-gallon oil drum cut in half.

Residence 5 was mapped in December 1987, while Household 2 was in residence at their rockshelter, Residence 4 (Table 3). The household did very little to prepare for the seasonal abandonment of Residence 5 and the residence continued to be used for storage during the winter occupation of Residence 4. Structure A (Figure 26) was used for storage in December, 1987, when I mapped it. Both the beds were covered with bedding and food supplies (which is frequently the case) and with cardboard boxes of clothing and other goods. Also, the jars used for fermenting and serving corn beer were stored upside-down close to the stove. This space is used for food processing and preparation when the residence is occupied. In this time of temporary abandonment, Structure B retained much of the same appearance it has during occupation, in part because this more open structure does not serve as the same kind of long-term storage area as does Structure A. The items in the northwest corner of Structure B are in long-term storage, however. In addition to the use of the two residential structures for storage at Residence 5, there is also the storage structure (Figure 13). Typically, storage structures contain not only the majority of stored foods, but additional cookware, valuable tools, and occasionally, clothing. Storage structures are some of the most private places for the Rarámuri. People are quite sensitive about having other people know what is inside them, and because storage per se was not the focus of my study, I did not inventory their contents.

Table 4 lists the materials recorded in Figure 26 for Structures A and B. In particular, the presence of the food processing and preparation facilities, and the presence of unusual items like the corn beer jars, violin, axe, and mallet should be noted. These objects are each expensive, either in terms of local production costs or cash value. If their use is not anticipated at the occupied residence, they are stored at the main residence.

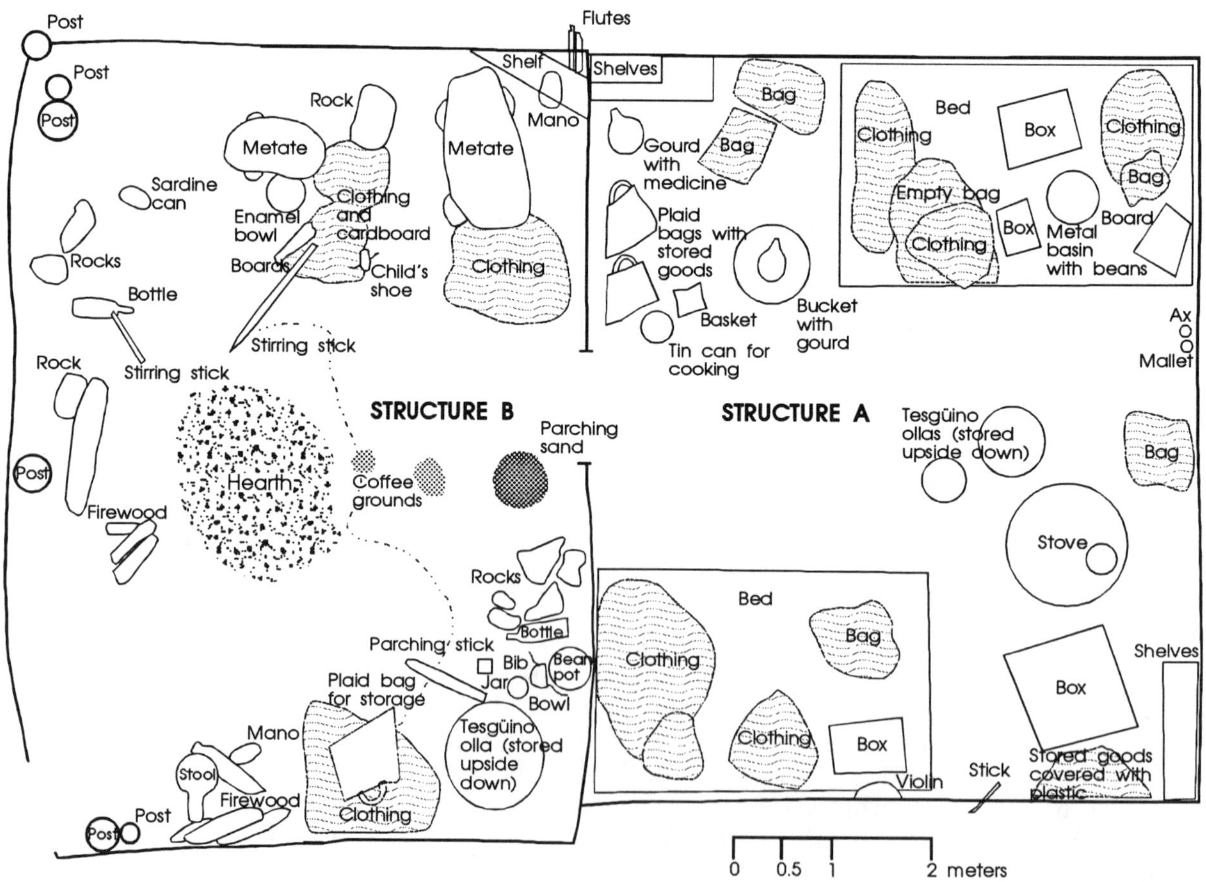

Figure 26. Interior of Residence 5 (originally published in Graham 1993:34, Figure 3.8).

Table 4. Items left inside structures A and B during abandonment of Residence 5 (see also Figure 26).

	Item No.	Name	Description
Structure A			
	1	Shelf	Shelves on wall holding various small packages and goods (no inventory)
	2	Bed	Two beds located in opposite corners of the room, with various items stored on and under them)
	3	Stove	Made from a 55-gallon drum cut in half
	4	Bedding	Various pieces of old clothing and blankets
	5	Box	Containing various items, including clothing, foodstuffs, and tools
	6	Misc.	Pile of items covered with plastic
	7	Clothing	In various stages of age and use
	8	Grain sack	Some of these contain maize or other foodstuffs, some are empty, others contain other items
	9	Stick	Use unknown; stored in wall of dwelling
	10	Violin	Hung on wall of dwelling
	11	Axe	
	12	Mallet	(Long-handled)
	13	Tin can	Empty
	14	Board	
	15	*Tesgüino* olla	Stored upside-down
	16	Bucket & gourd	Plastic bucket with gourd container inside
	17	Gourd	Contains medicine bottles
	18	Plaid bag	Contains squash/pumpkin seeds
	19	Plaid bag	Contents unknown
	20	Maize	Approximately 20 ears on bed
	21	Basin	Small, deep, made of metal, containing beans
	22	Cooking can	Fire-blackened, with wire handle
Structure B			
	23	Metate	In upright position for use (2)
	24	Mano(2)	
	25	Padding	Clothing and cardboard used as cushion while grinding
	26	Stirring stick	Used in cooking (2)
	27	Parching stick	For stirring maize while parching
	28	Parching sand	Prevents parching maize from scorching
	29	Sardine can	
	30	Plastic bottle	Saritan brand cooking oil (empty)
	31	Glass bottle	Tequila bottle (empty)
	32	Enamel bowl	
	33	Bean pot	Ceramic
	34	Bowl	Ceramic
	35	*Tesgüino* olla	Stored upside-down
	36	Plastic jar	Petroleum jelly (empty)
	37	Child's bib	
	38	Plaid bag	Contains clay
	39	Clothing	Old skirt (used for padding?)
	40	Shoe	Child's tennis shoe
	41	Flute	Hand-made (2)
	42	Stool	Hand-made from stump
	43	Firewood	
	44	Coffee grounds	Dumped on dwelling floor
	45	Rocks	Some are probably used for sitting, while others are used for propping up containers cooking in the hearth

The home base serves as a major storage facility both when the residence is occupied and when it is not. People spend a large amount of their "residential time" at these places, and perform a number of different kinds of activities (Camilli 1983). The main residence contains both expensive and infrequently used items that are then carried to each habitation as it is occupied. The artifact assemblages at these sites include some curated tools brought to and from the home base on residential moves, as well as a generalized site furniture assemblage that remains at all residences, regardless of the duration of occupation.

Variability in artifact assemblages regarding certain artifact classes may be one way to distinguish between the main and other residences. As I have already said, however, these residences do not necessarily retain the same function throughout the course of their occupation. If a habitation is abandoned as a main residence, and periodically reoccupied during the agricultural cycle, its assemblage will acquire characteristics of the agricultural residence. Permanent abandonment ends any residence's occupational history (until the site is reused). In the case of Residence 2 (Table 3), the household dismantled the dwelling and removed the construction elements and all useful tools and equipment to a completely new location. Only some structural foundations and trash were left. In contrast, the large "assemblage" listed in Table 4 is an inventory of the belongings left at a temporarily abandoned residence. When discussing the sites in terms of artifacts commonly recovered in archaeological assemblages, fewer distinctions can be made between the trash found at the main residence and trash found at other residences.

Residences Occupied During the Agricultural Cycle

A residence occupied during the agricultural cycle is inhabited for substantially less time each year than is the main residence. The Rarámuri move to these sites in order to carry out a variety of agricultural and social activities. Households periodically occupy these houses while tending to their own fields and also while assisting friends and relatives in that community (members of their labor network). As implied, frequently, these residences are located in a different community than that of the main residence.

The agricultural cycle residence holds an assemblage of site furniture, left in anticipation of return and reoccupation (Graham 1993 discusses Rarámuri abandonment and reoccupation at length). This assemblage includes cooking tools and vessels, and food processing facilities—i.e., the manos and metates. In many respects, this assemblage duplicates that found at the main residence. More valuable and transportable tools and utensils are stored at the main residence rather than cached at these habitations. For example, the molino is a mechanical grinder used most commonly for grinding processed maize to make tortillas and tamales; an activity that occurs almost daily (Figure 27). It is an expensive and heavy (although easily movable) item. The molino is carried to each residence as it is occupied throughout the course of the year. The molino shown in Figure 27 was brought to Basíhuare when the household made its temporary move there to attend the Holy Week festivities. Important facilities duplicated between the main and agricultural cycle residences include storage facilities, pens, corrals and barns for animals, drying structures, windbreaks, and specialized activity areas.

Ideally, artifact assemblages at periodically occupied residences, like those inhabited during the agricultural cycle, would be distinctive. These residences are more activity specific, with shorter periods of occupation, and therefore have fewer activities and kinds of activities represented than does the main residence. In any given year, reoccupation may be as a main residence or one occupied during the agricultural cycle. This results in a blurring of the occupational sequences we might expect in the archaeological record, so that Rarámuri residential assemblages probably are not sufficiently distinct to be discerned archaeologically. Moreover, as will be discussed in greater detail below, because the activity structure at Rarámuri residences remains relatively constant throughout the year, assemblages do not reflect differences based on season or length of occupation.

One occasionally sees evidence of another form of occupation in or near agricultural land in the study area. These are ephemeral structures providing shelter from the weather. These structures are not occupied to protect the fields from pests, but rather to watch over the sheep and goats corralled on the fields at night to manure the land (W. Merrill, personal communication, 1988). Members of a household will sleep at these shelters to protect their animals from predators (W. Merrill, personal communication, 1988), and these structures frequently provide sleeping space and a hearth for warmth and heating up snacks. Most meal preparation and all other daily activities are done at a residence. A shelter, hearth, and perhaps sleeping boards may be the only facilities at these sites, with no artifact assemblage in evidence. Residence 7 of Household 3 is an example of this type of occupation. There are variations on the use of these structures as living space; in 1983, one household in Rejogochi occupied a winter residence as its main residence, and during the summer slept close to the field where its animals were corralled. It had built a shelter near its storage structure as protection from the rain, enclosing a hearth, metate, and sleeping boards. By 1984, the household had built a residence at this location, occupying it year-round (R. Hard, personal communication, 1984).

Cold Weather Residences

The main residence is a warm weather dwelling. The Rarámuri occupy cold weather residences for periods varying from a few weeks to months. These residences are empty for the major part of the year. The Rarámuri inhabit winter residences for comfort and protection from the weather rather than in pursuit of agricultural activities. This occupation is for the benefit of both people and animals.

Compared to those occupied during the agricultural cycle, winter residences are season-specific, and occupied for relatively shorter periods of time. I expect that the Rarámuri carry out a smaller

Figure 27. Woman using a molino to grind corn. After boiling the corn with lime for several hours, the Rarámuri woman rinses it with fresh water, and does a "first grind" using the molino. She grinds the corn directly into a basket, as shown. The woman will then grind this coarse meal more finely using a mano and metate before mixing it with water and shaping the dough with her hands. In the final stage of preparation, she will cook the tortillas on a griddle. This woman is one of the temporary residents of the mule station in Basíhuare, one of the places the Rarámuri occupy during their short-term residential moves to participate in religious celebrations. The household brought this molino from their main residence in Rejogochi.

range of possible activities at these residences than they do at warm weather residences. Assemblages at these residences should contrast more with assemblages of warm weather habitations than the main and agricultural cycle residences will contrast with each other. Of course, assemblage variability alone would be inadequate for identifying site function. While the Rarámuri incorporate seasonal wild foods into their diet, during the winter only stored foods are available. Many of the religious ceremonies that include feasts of meat stews occur during cold weather. Under these cooler temperatures, these special foods can be stored longer than would be possible later in the year.

However, material assemblages at cold weather residences look very much like those of all other residences since subsistence activities occurring at the residential sites of agriculturalists do not vary substantially from season to season. Regardless of time of year, the basic food processing and preparation activities carried out at a residence are the same, because the diet remains unchanged: dried corn and beans, supplemented with some squash, seasonal wild foods, meat and some purchased foods.

One significant way in which winter residences may differ from other residences is in the spatial constraints inherent in their location. Because winter residences are occupied for comfort and protection, they frequently occur in rockshelters, caves, or other sheltered places. Less emphasis is placed on the proximity of these habitations to fields. Rockshelters and caves in particular have definite spatial constraints not found at other residences, having rigid back walls, sides, and overhangs. Frequently the mouth of the shelter opens onto a steep slope. This area is often built up and used as outdoor activity space. In general, the natural layout of rockshelters dictates some of the design of space use.

Residence 4 (Figure 28, Table 3) is a rockshelter occupied by Household 2 during the winters of 1986/87 and 1987/88. This rockshelter has been used by other household occupants over the last ten years (W. Merrill, personal communication, 1988; Hard and Merrill 1983; Graham, fieldnotes, 1987-88). Residence 4 is about a ten-minute walk from Residence 5 where the majority of its inhabitants live during the warmer part of the year. The household obtains its water from a spring on the valley floor when living at Residence 4. If there has been sufficient precipitation during the year, water from a nearby pothole is also available.

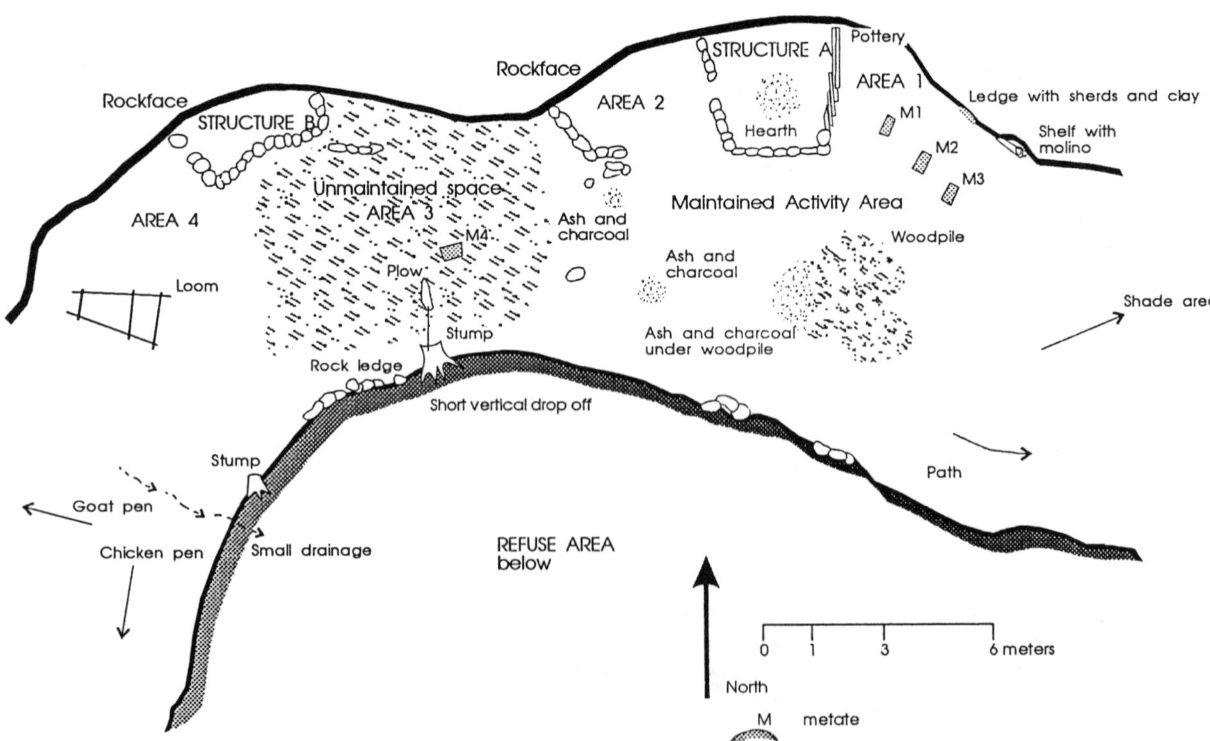

Figure 28. Residence 4.

Residence 4 has two walled structures, (Figure 28, Structures A and B). Structure A has an interior hearth where much of the cooking for the entire household was done. The structure has partial walls of stone on two sides, and logs on a third. The back wall and "roof" of the structure are formed by the rockshelter; however, in this area the rockshelter is shallow, and several planks have been set on supports to serve as a partial roof that deflects runoff from rains.

Structure B, located to the west of Structure A, is roofed completely by the rockshelter, and all three walls are stone blocks. Walls in Residence 4 and elsewhere are about chest-high. Walls in both structures end more than a meter below the roof. Other residences in rockshelters have even greater open spaces between the walls and the overhang of the roof. Householders find this necessary for light and ventilation, although a great deal of heat escapes in this way.

The household used both Structures A and B as sleeping areas. Structure A and two extramural areas surrounding it were used for many of the household's daily activities. Area 1 is in front of the rockface east of Structure A. Here, immediately east of the log wall of Structure A, Tenchi kept her pottery-making supplies, her newly fired pots and greenware, and a few of the household's commonly used cookware. These objects were stored on a narrow ledge. When making ceramics, Tenchi worked in the activity space outside the southeast corner of Structure A, probably because the area was both sunny and protected from the wind. The pots were fired at hearths shown on Figure 28 as the ash and charcoal areas in the Maintained Activity Area. (The household also made corn beer at the largest of these outdoor hearths.)

Also within Area 1 are three metates (east of the area where Tenchi kept her ceramic materials). Beyond these is a shelf for the molino. The metates are at least partially protected by the overhang, although the molino is not. I presume that in very cold weather the women would move these grinding facilities inside one of the structures, but for most of the winter, outdoor space was used. Two of the metates remained at the rockshelter during 1987. When the household abandoned this residence in the spring, it took the third metate and the molino.

In early 1987, Area 2 was a low-walled sleeping and activity area to the west of Structure A. Its walls were only one or two courses of stone. Bedding and large *tesgüino* jars were stored here, and it was both a sleeping and an activity area. In late 1987 and early 1988, Household 2 took boards from Residence 7 and constructed a shelter in this area. Toward early spring, 1988, the household dismantled the shelter, and reconstructed Residence 7.

Area 3 is the unmaintained activity area between Area 2 and Structure B. Most activities occurred in the Maintained Activity Area in front (south) of Structure A. Area 3 is used only for a few activities. Area 4, where the loom is set up, west of Structure B, is also not highly maintained. Weaving itself does not generate large amounts of debris, and few other tasks were performed in this area, so the amount of trash in the area, and the need for maintenance, is minimal. Both Tenchi and Jesusita used the loom during the 1986/87 occupation of Residence 4. It was not left at the residence when the rockshelter was abandoned at the end of the winter.

The goat and chicken pens are to the west of the main area of the residence (not shown in Figure 28). Upon abandonment these small structures were left relatively undisturbed. During the 1986/87 occupation, the chicken pen was at ground level. For the winter of 1987/88, it was rebuilt as an elevated structure rather than a pen.

In many cases, the lines of ownership and/or the annual occupants of cold weather residences are not as well-established as with other types of residences. Residence 4, for example, was occupied by Household 10 (see below) in 1979, Household 5 in 1983, and Household 2 in 1987 and 1988. Because over the years rockshelters in Rejogochi may be used by different households, few goods are cached at these residences. Nevertheless, assemblages of site furniture were observed cached at these residences. These always included manos and metates, and a few baskets, ceramic bowls and jars.

Storage facilities are infrequent at winter residences, particularly rockshelters. This is not surprising; the storage facility is the most important structure belonging to a Rarámuri. If a household is not firmly established at a winter residence (as in the cases of Residences 4 and 9, where occupants of the residence have varied over the years), there is less investment in facilities for year-round storage there.

During occupation of the winter residence, the main residence remains a major resource (Graham 1993); visits for supplies, food, and special utensils are common, especially if the distance between the main and winter residences is not great. In areas where cold weather residences are at some distance from the main residence, assemblage content at the winter dwelling may be more complete, similar to that found at the main residence, than is the case in Rejogochi (e.g., Champion 1962:106).

Table 5 lists items left at Residence 4 during the period of its abandonment in 1987. Household 2 had moved down to Residence 5 by mid-February 1987. A comparison of the artifacts recorded in late February 1987, and those documented seven months later in September show a difference in both number and kind of artifacts. In this case, abandonment is more a process than an event; items are gradually taken away for use at the household's main residences.

Tenchi made a number of pots while living at Residence 4. These pots were cached at the rockshelter only temporarily. Valencio was named Generál (General) for Holy Week in 1987. This position is an important one, requiring, among other things, that the household provide beer and food to all the participants of the celebrations (Merrill n.d. discusses the roles and responsibilities of this official in the study area; see also Kennedy and Lopez 1981). The ceramics recorded here were probably made in anticipation of the upcoming fiestas.

Temporary Residences

The community of Basíhuare is a religious-administrative center typically occupied by Rarámuri only during religious celebrations. It is located about an hour's walk from Rejogochi. The present study does not include in-depth descriptions, observations, or histories of sites used for temporary residences during these religious festivities, but a brief discussion is included here because these occupations are an important aspect of the Rarámuri lifeway and settlement pattern.

At one time Basíhuare was a Rarámuri community. Its church is probably of Franciscan origin, and according to Merrill's estimate, was built between 1828 and 1859 (Merrill 1988a:37). The community is now occupied predominantly by mestizos. The Rarámuri come to make purchases at the stores here, work at the sawmill, carry out government business, attend mass on Sundays, and participate in religious celebrations. As in Samachique (Bennett and Zingg 1935:184), a number of abandoned dwellings and rockshelters are maintained in anticipation of reoccupation during the religious events. An old mule station, built in 1907 as a way station on a road built for bringing silver up to Creel from Batopilas, is also used, both as a locus of religious activities and a temporary residence for numerous households. Because housing during religious festivities is fairly limited and households share social duties and responsibilities, two or more households often will share residential space. The mule station served as one Rarámuri household's main residence during most of my fieldwork.

Few of the rockshelters and abandoned houses in Basíhuare are suitable for lengthy occupations. At one time some rockshelters and, of course, the abandoned houses were used as more permanent residences; use of other rockshelters has always been limited to short term habitations. Rockshelters may be limited to temporary use when they receive too much or too little sunlight, heat, or wind. Others are too close to the perennial stream. At best, these are perpetually damp in places; at worst, they are subject to seasonal flooding. The abandoned houses are frequently unroofed, having only low walls still standing, and offering minimal shelter and privacy.

With a bit of modification and some maintenance, all become viable temporary residences within a matter of hours. Branching posts are set up to serve as closets and shelves for clothing and food, brush and small trees are cut and strategically placed as windbreaks and shades, and damp places are covered with sleeping boards, or used as storage areas for ceramic and waterproof containers. Households bring food supplies and cooking equipment with them from their more permanent homes (for example, the molino in Figure 27).

A few metates and large jars for *tesgüino* are cached at these temporary residences, but the common household assemblages are absent from these residences when they are abandoned for the rest of the year. Generally, there is no very good place to store ceramics or utensils. There is also no guarantee that a household will occupy the same temporary residence the next time it stays in Basíhuare, because it is outside the agricultural system of most Rarámuri in the pueblo, and households apparently do not lay a property claim to these places. Basíhuare also sees a relatively large amount of traffic—Rarámuri who assemble each Sunday for mass, other Mexicans who live permanently in Basíhuare, and incidental travelers. "Outsiders," both mestizos and Rarámuri who are not well known, are typically characterized as thieves and not trusted. It may be considered unsafe to leave important or valuable items under these conditions.

Table 5. Residence 4 during 1987 abandonment (see also Figure 28).

Area/Date Recorded	Structural Elements or Features	Items Abandoned and Comments
Structure A		
2-21-87	Doorway is blocked with a of tin (used for cooking tortillas?) held upright by a few rocks. A large rock (from Area 2?) has been set up to block the doorway. A smoothed pole (used for carrying water) is also along west wall. All these items serve to block the entrance to Structure A.	Roof is made up of four planks, the outer two longer than the middle <u>Inside Structure</u> In doorway: Tin used for cooking tortillas Hung at door: Grocery bag of clothes and cloth shawl Hat Along middle of north wall: Two blankets (with something underneath them?) Along east wall: Large shallow ceramic bowl used in pottery manufacture, together with one broken piece of a pot on top of it, and also a second piece of broken pot Along middle of south wall: Old basket In southeast corner: Bowl with water Scattered on the floor: Rope Wooden corn parching stick Part of small animal trap(?) On post in southwest corner: Woven hat? <u>Niche in Rockface</u> 2 small baskets 1 deeper basket (small) 1 battery 1 small medicine bottle cloth <u>Roof</u> Old pieces of squash (or rinds) 1 small bean pot 1 medium-sized basket 1 very large drinking gourd 1 cardboard box with rolled-up clothing inside
9-20-87	West wall fallen inside structure. Two short poles are in the area west of the structure. East wall partially collapsed; top log (sixth from bottom) has fallen out. Two other posts have fallen out. Northeasternmost meter of this east wall has the base log still in place. Activity area in front of structure is grassy and irregular, but with few artifacts (small fragments of objects).	<u>Inside Structure</u> Remnants of old shirt Shirt (whole or nearly whole) Along north wall, near northwest corner: Broom (made of grass) Near north wall: Poles for carrying buckets of water By northeast post: Broken pot base with small amount of sand inside (probably for use in parching corn) <u>Niche in Rockface</u> Shirt (rolled up) Old blouse (child's?) <u>Roof</u> On westernmost plank, protected by the overhang: 1 small pot (the size used for cooking greens; relatively new?) 1 small, thin mano 1 old, worn mano 1 wooden weaving tool(?)

Continued on next page

Table 5—*Continued*

Area/Date Recorded	Structural Elements or Features	Items Abandoned and Comments
Structure B		
2-21-87	Doorway is blocked by various poles and two planks (from Area 4?). One rock at the bottom of the wall near the rockface has been dislodged, and there is a hole.	Inside Structure 5 large and medium cooking pots (possibly also for *tesgüino*) 2 unfired pots 1 Fanta soda bottle 1 five-pound coffee can with a cloth handle 1 old child's skirt 1 small, old plastic grocery bag 'other clay things'—presumably for ceramic manufacture
9-20-87	Logs blocking doorway have fallen in and out of the structure. Only two are still standing.	Inside Structure (Pottery manufacturing items and ceramics gone) 1 long fire-blackened pole 1 old basket with large pieces of the base of a large cooking pot 2 old shirts 1 Fanta soda bottle
Area 1 (Grinding and Ceramic Manufacturing Area)		
2-21-87		(All the pottery here during occupation is gone, either off the site, or in Structures A and B. Metate #1 [easternmost] is also gone.) Various pieces of cloth, rope, and grain sack 1 base of a basket 3 *ŕekótzara* (hammerstones for pecking and shaping metates and manos) 1 wooden ball Niche in Rock Some clay and old pieces of pottery (for ceramic manufacture—clay and temper) Shelf for Molino (The molino is gone, but the board used as a shelf remains) 1 unfired bowl 1 broom (of grass) to sweep off metates 3 manos forked sticks, small (use unknown)
9-20-87		(Rocks to support missing metate are gone, possibly scattered about. No *ŕekótzara* noted.) Shelf for Molino (Empty except for a small chipped gourd [used in pottery manufacture] at edge of shelf.)
Area 2 (Low-walled Sleeping and Activity Area)		
2-21-87	The front low rock wall has been removed completely. The fallen rock wall is gone except for a single rock. At least part of this rock wall has been used to block the entrance of Structure A, and possibly of Structure B.	(The ollas in this area are now gone—some are stored in Structure B, others presumably are at the residence currently occupied [Residence 5]. The bedding from this area is also gone.) 1 piece old grain sack

Continued on next page

Table 5—*Continued*

Area/Date Recorded	Structural Elements or Features	Items Abandoned and Comments
9-20-87	Area is defined only by three large rocks at west side, and one large rock on the south side, forming the southwest corner of the area.	None noted
Area 3 (Unmaintained Activity Area)		
2-21-87		(Long metal rod [used for excavating in rock, among other things] no longer here) Metate #4 Along north wall, on slope Clay (possibly for ceramic manufacture) Toy with wheels 'Sticks' (possibly unidentified tools or utensils)
9-29-87	Area is relatively clear, with some rock and other debris that has washed down from the overhang above.	Metate (upright) Notched post (leaning against rock face; this is a new item)
Area 4 (Weaving Area)		
2-21-87	The plank that acted as a shade resting on upright posts is now gone, probably used to block the entrance of Structure B.	(Loom is gone.)
9-20-87	The two upright posts of the shade remain, but nothing else. There has been heavy erosion from here into Area 3.	None noted
Goat Pen		
2-21-87	The 'door' of the pen is open.	None noted
9-20-87	The pen is relatively clean inside, and the structure is in good condition.	None noted
Chicken pen		
2-21-87	Unmodified from period of use.	None noted
9-20-87	Only the frame of the pen remains; it now resembles a loom, except the logs are not notched. The other logs used for the pen are no longer at this residence	None noted
Woodpile		
2-21-87		(Most of the wood is gone.) Flute
9-20-87	Heavy erosion in this area.	None noted

Continued on next page

Table 5—*Continued*

Area/Date Recorded	Structural Elements or Features	Items Abandoned and Comments
Beneath Low Rock Ledge		
2-21-87	(Small overhang is used as a cache area)	Rock used in the Maintained Activity Area for rinsing corn is here; perhaps some other items as well, but I could not see clearly, nor could I look more closely without permission.
9-20-87		(No ceramics noted) Rock used for rinsing corn is here
Shade Area		
2-21-87	The remains of the brush used to shade this area the day I was mapping is still here.	Wooden ball Belt with buckle Rock used for sitting

Activities at Rarámuri Residences

The arrangement of structures and facilities at a residence provides a framework within which to perform activities. The nature of these activities and their placement at the residence both influence and are influenced by the basic layout of the site. This section looks at the placement of tasks in detail. Reviewing spatial organization of tasks in a discrete section highlights the role of the placement of activities and facilities in any subsequent organization and use of space at any given residence.

Food Processing and Preparation

In this study, food processing is any of the initial phases in a sequence of breaking down foodstuffs, while food preparation encompasses the final stages of foodstuffs reaching their edible consumable form. For example, the steps of food processing to make the dough from which tortillas and tamales are made include shelling corn, boiling the kernels with lime for several hours, then rinsing and grinding the cooked maize. Forming the dough into tortillas or tamales and then cooking these are examples of food preparation. Food processing and preparation activities are among the most common activities occurring on residential sites. Shelling corn, cleaning beans and seasonal greens, parching, boiling, grinding, and supplemental cooking of some foods are all food processing and preparation tasks. In Rejogochi, most food preparation takes place at residential sites and represents a major class of daily activities occurring at these places.

Food Preparation Activities

Mares Trias (1982) has described a number of Rarámuri foods and their means of preparation (see also Pennington 1963). Boiling is the principal method of cooking. Baking and parching are the other common techniques. More implements are associated with boiling foods than with other kinds of cooking. All boiled foods must be contained while cooking, tortillas are baked on a griddle, and corn is parched in a special vessel, but meat and potatoes may be roasted in the coals or boiled. On the rare occasions that one household has a relatively large amount of meat to process for itself, it may cut the meat into long thin strips and hang them to dry and be smoked. Fish are also dried and stored. Before eating the dried fish, they are warmed on the coals of the hearth.

The most common methods of processing dried corn begin with shelling the kernels, then parching or boiling them. For green corn, the kernels are simply cut off. They are then ground into meal (flour) using either the molino or metate. Rarámuri boil their corn with lime to make the nutrients in the maize more digestible. When it is ground, it is used in making tortillas or tamales; otherwise, it is called posole. Parched corn is prepared by putting it into a special container with some sand to prevent scorching. The vessel is heated on the hearth, and the corn stirred constantly to prevent burning. After the kernels pop, they are ground. If ground alone (ground dry), a fine meal is produced (Figure 29). Ground with water, the parched corn becomes a paste. Both forms of parched corn are mixed with water and drunk.

One major food processing and preparation

event not commonly carried out at residences in Rejogochi is the baking of agave in roasting pits. Agave is a seasonal food, and is by no means a major supplement to the diet in this part of the Sierra. When residents of Rejogochi prepare this food, they generally do so at or near the place where the agave grows. Pennington (1963:129-130) discusses transport of the agave back to the residence to be roasted nearby, and Robert J. Hard (personal communication, 1983) documented roasting pits near Rarámuri residences in the barrancas, but away from the Maintained Activity Areas and dwellings.

For drinking parties and fiestas, a household processes large quantities of corn. They may butcher one or more animals for fiestas (usually goats, but occasionally a cow), and the hosting residence is inundated with people. Agricultural or other tasks are commonly associated with these occasions, and while the men are working in the fields, or some distance from the residence, the women are working at the dwelling, preparing food, cleaning wool and preparing it for spinning, or doing other household chores.

These are very much social occasions, and frequently a woman will ask friends or relatives to help with the preparations, which begin a couple of days in advance of the work party. During the work party and fiestas, visiting women also assist in processing food for the host household. In some cases, visitors will also take some of this food home with them; this is especially true when the hosts provide meat. Thus, although the activities of food processing and preparation remain the same, the locations of the activities may change to accommodate the increased number of adult women involved in preparing meals. This change also allows greater interaction and visiting between women involved

Figure 29. Woman grinding parched corn using a metate. The corn meal falls from the metate into a wooden trough. A second woman, in the upper right of the photograph, also is grinding on a metate. The household is hosting a religious fiesta at this residence. The metates have been placed in the Maintained Activity Area so that the women invited to the fiesta can socialize and work at the same time. Behind and to the left of the woman in the foreground, a girl prepares to carry her younger sibling in the traditional manner.

at separate tasks, and to accommodate the large quantities of food in the various stages of production (Figure 30, see also Figure 2).

Based on his fieldwork in Inápuchi, Kennedy states "...it is no exaggeration to estimate that the average Tarahumara spends at least 100 days per year directly concerned with *tesgüino* and much of this time under its influence or aftereffects" (1963:635). This generalization is fairly accurate for the Rarámuri of Rejogochi as well. Although the work party that includes drinking home-brewed beer is not an everyday occurrence, it is a common one. Any discussion of activity structure at Rarámuri residences should consider it in association with work and drinking parties. Households occupying residences in a community during the agricultural cycle will also host drinking parties, again implying a the full range of activities and artifact assemblages associated with these events at the main residence.

Animal processing is also constrained by activity structure and areas. The killing and butchering of animals for a fiesta or work party is done within full view of the community. For example, two separate fiestas at Residence 11 (Figure 31) involved the sacrificing of animals. For the first event, a young cow was killed on the opposite bank of the arroyo from the residence. Sorpino and his assistants did the initial disarticulation at that location before bringing the meat over to the residence. The meat was stored temporarily on the storage structure, and subsequent butchering was carried out there before the final processing for the traditional stew. The actual cooking of the meat

Figure 30. Women preparing foods for a religious fiesta. The Maintained Activity Area, where the women sit, is littered with corn husks used to wrap tamales. The large metal cooking pot contains the meat broth, cooked overnight, and now cooling. Once cool, the women will strip the meat off the bones, and later they will distribute portions of meat and broth to each household. The hearth where the cooking pot sits will be broken up at the end of the fiesta (or sooner, if this space is used for dancing). The large rocks in the hearth, used to support the pot during cooking, will be stored. The ashes will be swept up and dumped in the Concentrated Refuse Area.

and bones took place at the hearth in the sheltered workspace between the two structures (the ash scatter in Figure 31 is the location of the hearth). For the second event, two goats were sacrificed in front of a cross set up for the occasion in the Maintained Activity Area to the west of the north corner of Structure B. One goat was then hung from this corner of Structure B for the primary butchering. The other was hung from the eastern corner of the same structure. Subsequent processing of the meat again took place by the storage structure, near the hearth in the sheltered activity area and at the hearth inside Structure A.

These sacrifices are an integral part of the celebration, and as such are done openly, in a special place. In the case of the two goats, for example, the place was defined by the location of the crosses, put up in the Maintained Activity Area specifically for this fiesta. During other events, the location will vary, although the setting (that is, the crosses and the public view of the event) will not.

Animals intended for private rather than communal consumption are dispatched quickly and quietly out of view of other households in the valley. Chickens and small rodents are processed at the hearth, either inside the dwelling or in the sheltered workspace. They are easily skinned and gutted at these locations because of their size, but concealment is also a factor. The processing of larger animals is also concealed. When Household 3 killed and butchered a pig, it did so while the majority of residents of Rejogochi were at a *tesgüinada* in the upper valley. Santa Rosa, assisted only by her eight-year-old son, dispatched the pig inside a nearby abandoned house now used as an animal pen so no one would observe them.

Food Preparation Facilities

In addition to containers and utensils, food processing and preparation activities require various facilities. Among the Rarámuri these include the hearth, the molino, and the metate for daily subsistence, as well as other facilities, for example roasting pits, for seasonal foods. Shelling corn, mixing boiled corn for tortillas, and other activi-

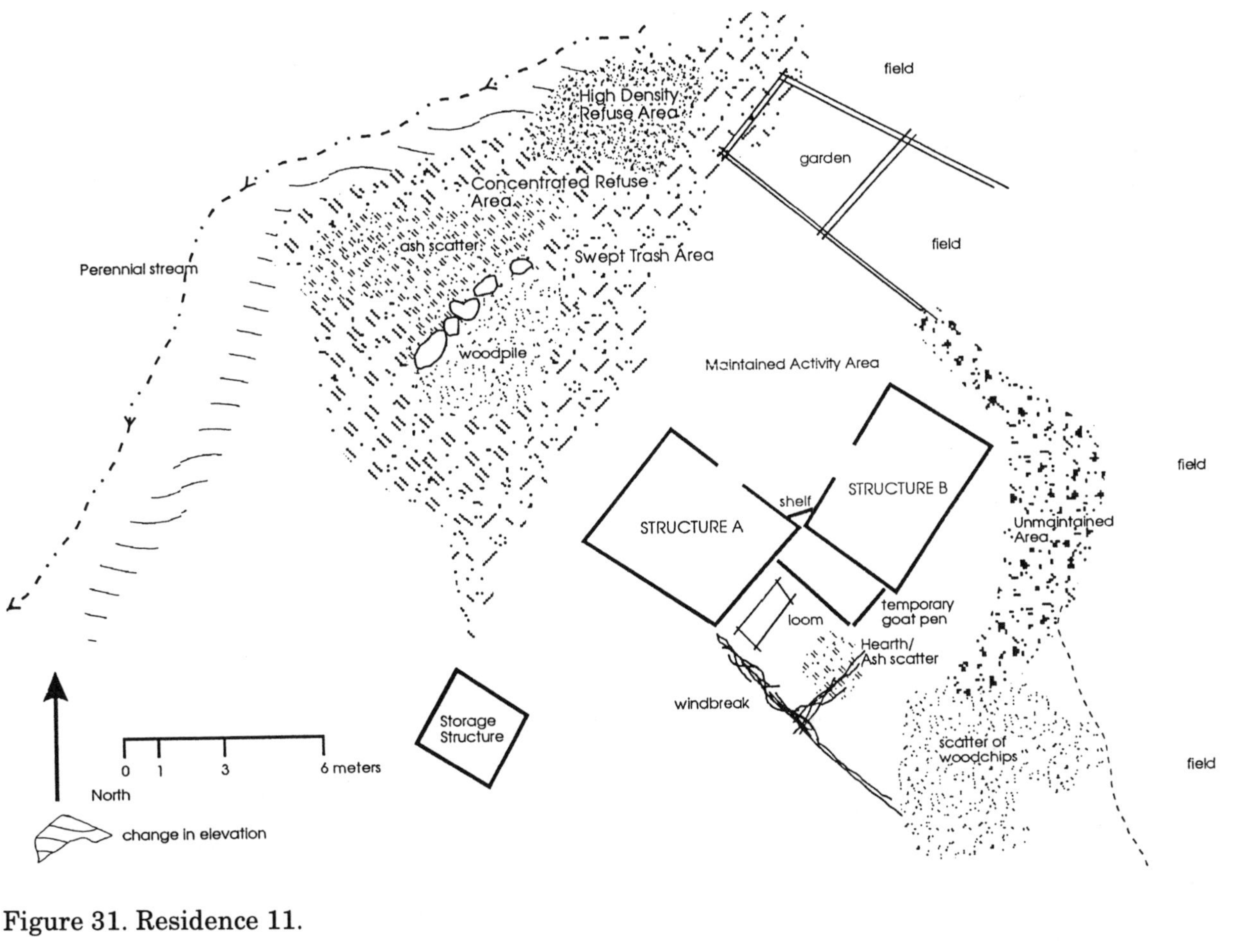

Figure 31. Residence 11.

ties not requiring a facility tend to occur anywhere within the dwelling, Maintained Activity Area, or sheltered activity space. When shelling corn, for example, a person may sit out of doors in the sun or shade, or on a bed near the interior hearth. Activities requiring facilities by definition occur at that facility. Nevertheless, an important aspect of Rarámuri residential site structure is that the placement of facilities remains relatively flexible within the dwelling and Maintained Activity Area. Metates, for example, are moved throughout the residence in response to various situations (for example, Figure 29). One facility not yet mentioned, yet possibly the least-often moved, is the rock or ledge that is used to rinse the boiled corn. Almost every residence has such a place, although when a large amount of corn is being processed for a fiesta, women and children rinse buckets of boiled corn at the perennial stream rather than hauling water to the residence for this purpose. This facility is important enough that the rock used at Residence 4 (Figure 28) was cached under a ledge when the rockshelter was abandoned in the spring.

Rarámuri facilities are not fixed in space. Various activity areas and facilities for Residence 11 (Figure 31) serve to illustrate this point. During my stay in Rejogochi, María Elena moved the molino at Residence 11 three times. I first recorded it in use on a shelf set on a post and wedged into the wall of Structure A. Later, it was moved outside to a shelf between the two structures (where the chickens sometimes roost). On occasion, it was put inside Structure A on a shelf in the southwest corner of the house. When the household moved to a rockshelter for part of the winter (Residence 9, Table 3), María Elena carried the molino up to this residence to use, as well.

During my first visits to Residence 11 I noted a single metate in Structure A (Figure 31). While preparing for a fiesta, María Elena put her second metate in Structure A. This arrangement lasted for several weeks, and was occasionally repeated throughout the period of my fieldwork (not always in the context of large work parties). When I left in the spring of 1988, both metates were being used in Structure B, but more commonly, María Elena had one metate in Structure A, while the second was in Structure B. When grinding corn, Rarámuri women typically do an initial grind on one metate, and use another metate (with a different mano) for a second, finer, grind. Thus, María Elena and other women in the community may move from one dwelling, or part of the residence, to another during the sequence of grinding. Moreover, other women, for example Irene of Household 8 used only a single metate for processing corn (Figure 32, see also Figure 21).

Santa Rosa (Household 3) used a number of different cooking areas at Residence 8 (Figure 33). At this residence, Structure A had a stove, and Structure B had a hearth. Throughout the year, various food processing and preparation tasks were carried out in Structure B which provided more light and space within which to work. Santa Rosa used the stove in Structure A during cold weather when it provided both a more comfortable place in which to work and a more efficient use of firewood for cooking and warming simultaneously. In the summer Structure A was used very little. Santa Rosa and Rafael moved the stove into Structure B, and used Structure A as a storage unit. During the course of the summer and fall, Santa Rosa also moved many of her cooking activities outside into the Maintained Activity Area where a fire set in a break in the rock outcrop conveniently provided two tiers of heating—for warming and for hot cooking. Hearths for large-scale cooking, for example to prepare meat stew and corn beer for a fiesta, were north of Structure A in unmaintained space, east of Structure A in the Maintained Activity Area, and at the hearth inside Structure B. The latter was used when the Rarámuri wanted to avoid contact with the Mexican soldiers who were active in the area for several days and nights, documenting the possession of firearms, and looking for marijuana fields.

Residences 12 (Figure 19) and 14 offer another example of the relative flexibility of space use and the placement of facilities. Here, Guadalupe (Household 6) and her married daughter Sarina (Household 7) shared the use of two metates between the two residences. At Residence 12 I had observed one metate inside the dwelling and the second outside. When I was mapping the residence, both metates were outside. At no time did I observe a metate indoors at Residence 14; rather, both metates (those observed from Residence 12) were left out of doors, at any of two or more locations within the Maintained Activity Area.

Thus cooking and grinding facilities are associated with several locations at the Rarámuri residence according to various constraints or situations including the weather, the number of workers present, and the foods being processed. Food processing and preparation activities not requiring a facility are less tied to any specific place, and those factors influencing the use of one place over another, for example, sheltered space during in-

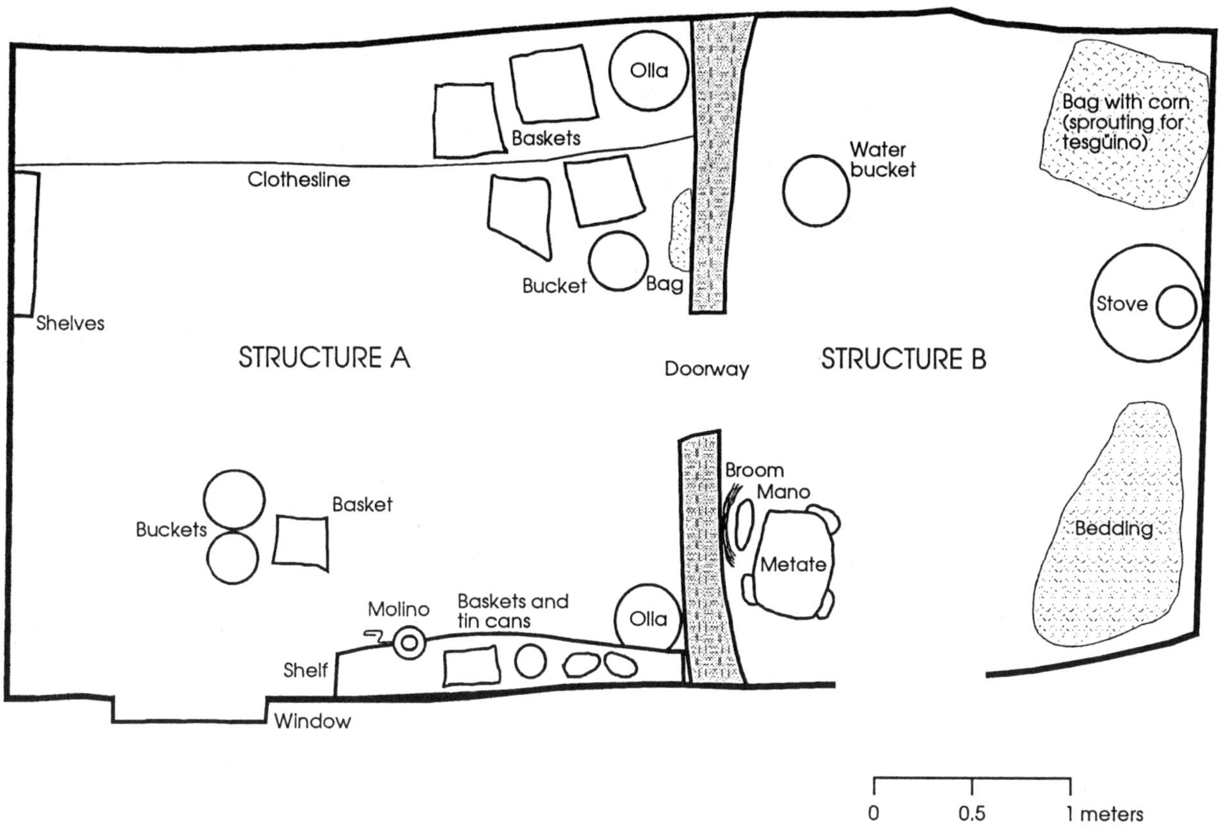

Figure 32. Interior of Residence 16 (from Robert J. Hard, with permission).

clement weather, are easily mitigated, often by simply erecting an ephemeral shelter.

Sleeping

Like food preparation, sleeping is a daily activity at residential sites, and usually requires a sleeping facility: a bed. Among the Rarámuri, beds are either made of planks, sometimes slightly elevated above the ground, or are merely bedding laid down on top of cardboard or some other insulator. These facilities are located inside dwellings, but in the case of two or more dwellings at a residence, the location may vary with season. Temperature, weatherproofing, and cooking activities influence where people will sleep.

During good weather (into early and late winter), sleeping facilities are also placed outside. In the walled activity area of Residence 18 (Figure 23), for example, Katilachi and Cornelio slept in a bed placed against the wall of Structure C (where the loom is shown) throughout much of the year. A second outdoor sleeping area east of Structure A was described above. In many cases, ephemeral structural elements are put up for additional protection.

Inside the dwelling, sleeping facilities are oriented toward the hearth or stove. Because hearths and stoves tend to be centrally located and the dwellings relatively small, this relationship is not unexpected. Positioning varies somewhat depending on whether it is a hearth or a stove.

The bed is used for more than sleeping, however (see also Binford 1983a:160-165). Because of its orientation and proximity to the hearth or stove and the smallness of the dwelling, the bed is frequently a focal point for food processing and preparation. A woman sitting on the edge of the bed can be in a comfortable position to tend the fire, parch corn, and monitor a pot of cooking beans. Babies propped up by the bedding and left to amuse themselves are within easy reach of their caretaker who, seated on the bed, is performing other tasks.

Craft Activities

Craft activities include weaving, ceramic manufacture, basketry, sewing, the making of musical instruments (for example, the drum, violin, and guitar) and other woodworking activities.

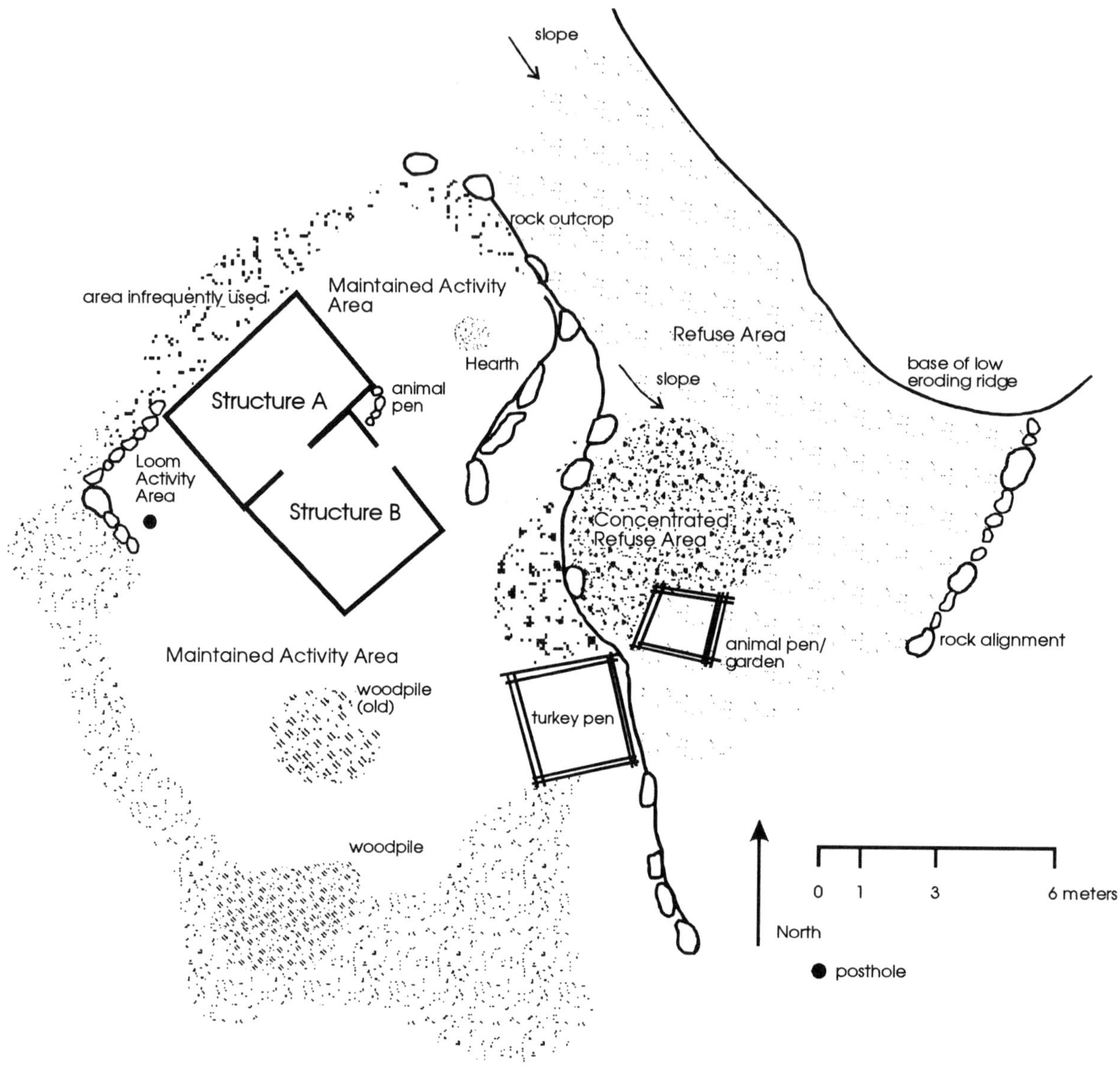

Figure 33. Residence 8.

The Rarámuri carry out some of these activities away from the residence. For example, in processing goat skins for drums, Rafael (Household 3) sat at a distance from his residence removing hair from the skin. Moreover, a number of old abandoned residential sites are now used as ceramic manufacturing loci. Located near good clay sources, these sites are also convenient to the potter's residence. Women cache their tools for ceramic production, as well as both fired and unfired pots there.

A great number (if not the majority) of craft activities occur at the residence, however. With the exceptions of weaving and ceramic manufacture, to be discussed below, these activities generally occur in the Maintained Activity Area. As with other kinds of activities, no particular place is set aside for crafts. Basketry, for example, is a time- and space-consuming task. Single baskets may take several hours to make. During their manufacture the work place is littered with the raw materials in various stages of reduction and baskets in assorted stages of production. Nevertheless, the Rarámuri make their baskets in the generalized activity area. If they spend several days making baskets, they will clear the area of debris each day.

Women frequently weave or engage in ceramic manufacture in sheltered but unroofed activity space (occasionally, men also become potters and weavers). Unlike the other activities mentioned, these need special facilities. Ceramic manufacture requires equipment including a pot base on which to place the pot, containers for clay, temper, and water. A place to make the pots, somewhere they can dry before firing, and a facility for firing are also necessary. Clay is often ground on the same metates used for food processing at residential sites. Elements of ceramic manufacture are stored when not in use, but during times of ceramic production, they are generally located in the sheltered but unroofed activity area, or at least in some space where they are easily accessible, and near or at the work place where the ceramics are made. Ceramics are fired in an outdoor hearth at the residence. When several items are ready for firing, the woman gathers fuel and the ceramics are all fired in one or two episodes. Single small- to medium-sized pots and bowls are often fired at hearths concurrently cooking corn beer and/or meat stew for the fiestas at which the vessels will be used (Pennington 1983:Figure 7, p. 283). It seemed to me that ceramics, especially bowls and the larger cooking and serving vessels, frequently are made to be used during large fiestas. Subsequently, they are used until they break, but bowls in particular are not replaced until preparations for the next fiesta begin.

Weaving also requires a special facility—the loom. As with most Rarámuri facilities, the loom is easily moved. Typically, a woman transports the same loom between cold weather and warm weather residences, something not done with metates. Although the illustrations of residences locate looms most often in sheltered activity areas (Figure 34, see also Figures 19, 23, 28, 31, and 33), the portability of the facility should be kept in mind. The majority of the residences were mapped in early spring, when the weather was still cool and a bit uncertain. Perhaps more than any other activity observed at Rarámuri residences, weaving is subject to the environmental constraints of the seasons. With the possible exception of Residence 17 (Figure 34), which was not visited during other periods of the year, the location of *all* looms illustrated at the various residences changed at least once.

Often we conceptualize specialized activity areas as places set apart to provide uninterrupted work space or time. While this is true, there is a danger in taking the notion too literally, at least in the case of the Rarámuri. For example, looms are usually set up in a special activity area that can be sheltered. The special activity area is also defined by its activity structure—few activities occur here (i.e., many classes of activities usually do not) and weaving either belts or blankets requires a number of days, if not months. This space nevertheless becomes the focal point of a number of social activities. When a woman is working in a specialized activity area, her children are usually "helping" her, or playing nearby. If a visitor comes, the women socialize while the weaver continues her project.

Socialization and Leisure

Visiting, leisure-time, child socialization, and other forms of socializing occur in the Maintained Activity Areas, within dwellings, and other social lifespace. It is customary to feed visitors, and food is often left in an area containing kitchen wares and utensils. I have been offered tortillas stored in one house to accompany beans stored in the second dwelling, or *pinole* that was in one dwelling with agave brought from a separate dwelling-qua-storage structure. A visitor is often invited to eat inside while the host remains outside. After eating, the visitor then joins the host wherever the latter is lounging or working.

The main areas for socialization are those where daily tasks are performed. During times of the day when the household is preparing meals, this may mean at a hearth inside or outside a dwelling. At other times, it means outdoors in the sun or shade depending on the season, against the side of a dwelling, in the main activity area, or perhaps at the loom.

Tool Maintenance and Manufacture

Rarámuri make, store, and maintain most of agricultural tools at their residences. Often such activities are performed in the context of socializing and relaxing, usually not requiring an extensive area or a period of uninterrupted work time. Making a new axe handle, sharpening the blade of an axe or knife, carving a new plow, or making rawhide rope are all tasks requiring but few tools, all of which are highly portable (an axe, a sharpening stone, a stick). These activities occur in the Maintained Activity Area or in the sheltered lifespace. During the tenure of my fieldwork, these activities were only infrequently observed. When questioned, the Rarámuri did not have explicit locations for performing any of these activities.

Of special note is the manufacture and maintenance of groundstone. Metates and manos are part of the site furniture at any occupied (and most unoccupied) residence. Here, maintenance refers to roughening of the metate or mano surface by pecking at it with a kind of hammerstone. A woman pecks her metate where it is set up. She may roughen the mano while at the metate, or carry it to another spot in the Maintained Activity Area to work on it.

I observed the manufacture of only a single mano. Santa Rosa was making this mano to grind sprouted corn for corn beer more efficiently as her other manos were too small and light to be effective. She sat inside her dwelling and pecked gently at a river cobble, shaping it. Although I left before the mano was finished, only one large flake (about 2 by 2 cm) was in evidence; a few smaller fragments were lying around, but a fine powder constituted the principal debris. A relative's baby daughter played around the woman's knees as she worked. Clearly it was a task requiring no special or uninterrupted work space, and virtually no cleanup at the end of the process (see also Bennett and Zingg 1935:79-80).

Animal Husbandry

Some activities relating to the caretaking of animals occur at residential sites. Animal pens are placed around the periphery of sites except when protecting the animals from the elements or predation is a major consideration. In the majority of rockshelters used as residences, goat and sheep

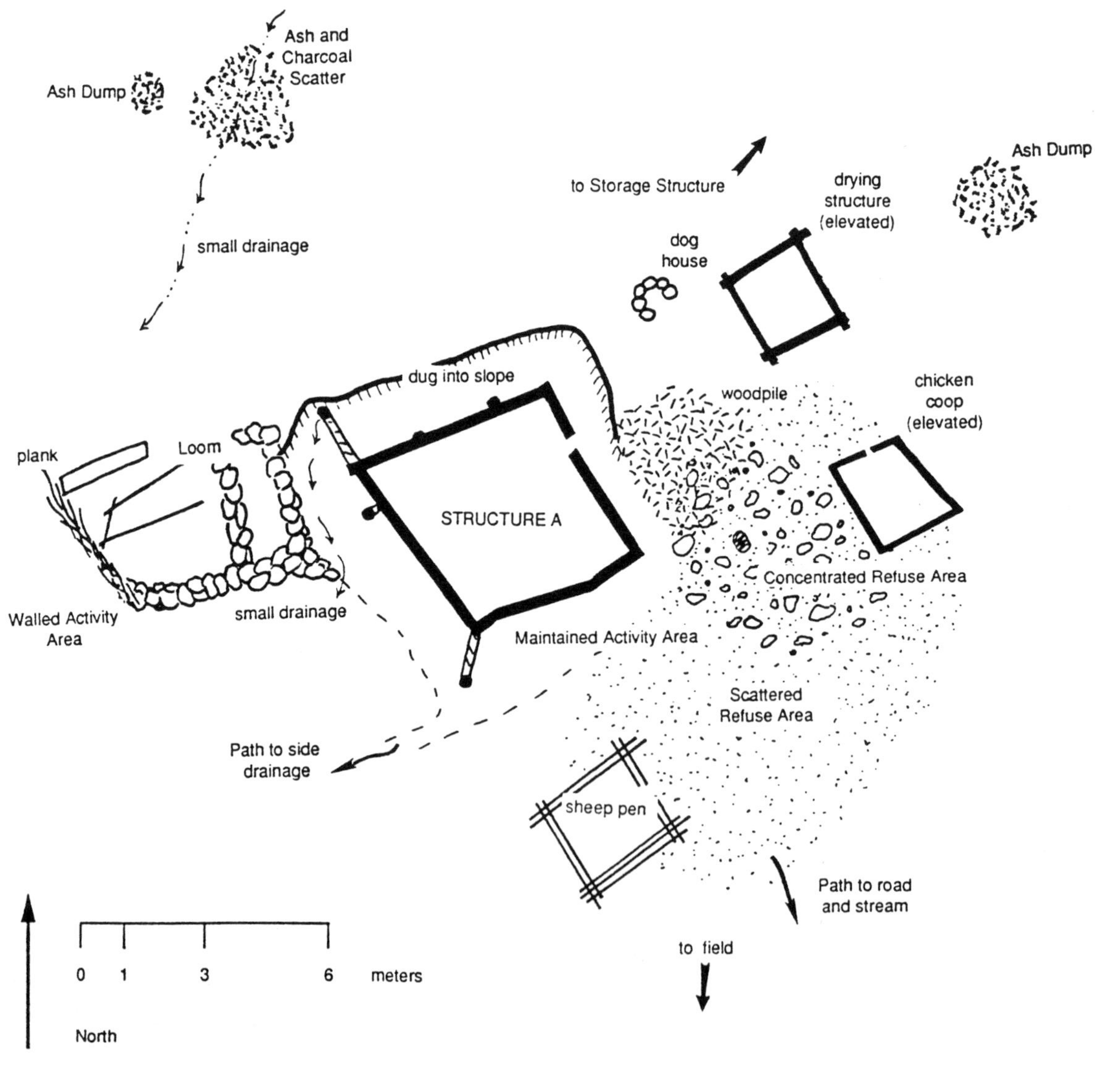

Figure 34. Residence 17.

pens adjoin dwelling space (see also the pen next to Structure C of Residence 18, Figure 23). Rarámuri often locate goat and sheep corrals at the periphery of the site and, after moving the corral, will use this space as gardens. This was the case with the sheep pen at Residence 17 (Figure 34). To my knowledge, the Rarámuri do not use special facilities during lambing season. I observed two episodes of branding cattle and horses. In one case, the men used a hearth at the edge of a field below the residence of the household hosting the branding party. In the second case, they used the general-purpose outdoor hearth at the residence, and the branding was done just beyond the edge of the Maintained Activity Area. Between two and four animals were branded in the latter case, while close to 15 were branded in the former one.

Agricultural Duties

In general, agricultural duties are carried out in the fields rather than at the residences. Men gather at the residence and bring their equipment for yoking oxen, but the rounding up of animals and subsequent plowing takes place in the fields. When planting, there is little preparation within the bounds of the residence. Planting sticks are left in the fields when workers take a break and return to the residence to eat or drink, although individuals store these implements at their residences during the rest of the year.

At harvest time agricultural activities do extend to the residence itself. The harvest is brought to the residence to be processed for storage, which entails putting the ears in drying bins and subsequently into the storage structure. If there is not enough room in the storage structure for the entire harvest, the Rarámuri will shell some of the corn and store the kernels in grain bags. Beans also require processing for storage. Rarámuri leave the beans to dry in the field or cut the plants and bring them to the residence, drying them on the roofs of dwellings or in nearby trees (Pennington 1963:63). The beans are subsequently threshed on rocky ground. In Rejogochi, the Rarámuri usually thresh along the banks of the perennial stream rather than at the residence.

Fodder is cut in early fall and stored in convenient trees near the residence or close to the field. Rarámuri sell some fodder to local mestizos, and also feed it to their animals when forage is scarce or the household is unable to take the animals out to graze.

Religious Ceremonies and Festivals

The Rarámuri perform a number of different kinds of religious activities at their residences. Ceremonies are often held in conjunction with, or as the reason for, *tesgüinadas* and fiestas. Merrill (1981, 1983, 1988a, 1988b) describes many of these events (see also Velasco Rivero 1987). At several kinds of festivals and different times of the year, dancing is a part of the occasion (Velasco Rivero 1987). In all cases, the Rarámuri use either the general activity area (during the agricultural season), or they will clear and level a portion of a field for dancing (Figure 35).

At the end of the ritual, the accoutrements used for the ceremonies are dismantled, and the space reverts to its previous use. Lumholtz (1902:171) refers to the dance patio as an element of any Rarámuri residence. In Rejogochi, the Maintained Activity Area serves as a dance area when necessary, but there is no special area given over to dancing or other religious events. It is likely that this was also the case during Lumholtz's time, as well, and areas he identified as dance patios were simply a part of the Maintained Activity Area.

Seasonal Variability

The activity structure at Rarámuri residences remains relatively constant throughout the year. While different season-specific activities do occur at the residences, the majority of activities are performed year-round. Throughout the year, the Rarámuri subsist primarily on their stored supplies of maize and beans, and therefore food processing and preparation techniques remain the same.

In late summer, before the harvest, depleted food stores are supplemented with other foodstuffs. Even before the end of the growing season, Rarámuri add seasonally available wild foods to their diet. Wild greens are available from early spring through summer. The Rarámuri collect and eat the pads and fruits of prickly pear cactus when they ripen in the spring. Other wild plants, including agave are also collected and consumed when available. Bye (1976, 1979, 1981) and Pennington (1963) have detailed accounts of the wild plants exploited by the Rarámuri. With the exception of the agave, which is baked in large pits, these wild foods do not require specialized facilities for processing or preparation. While evidence of wild foodstuffs exists in trash areas at Rarámuri sites, seasonal differences in food processing and preparation activities and activity areas are not distinguishable.

A number of activities at residential sites are season-specific. The drums used in the Holy Week ceremonies are made in the spring; they can only be played between February 2 and Easter (Merrill n.d.), and are most frequently manufactured at that time. This seasonal activity leaves behind little evidence—some wood shavings and bits of thong, leather, and ocher, perhaps. Religious events occur seasonally in coordination with the Catholic religious calendar (Velasco Rivero 1987). The community of Rejogochi is not a religious center, and here religious celebrations take place in the Maintained Activity Area of a residence, or in the surrounding field. Within a few weeks after their creation and use, these specialized activity places become invisible in the general use of the residence and area around it.

The activity structure of residential sites does not change at different times of the year in such a way as to be evidenced in the arrangement of residential elements in space. But the *placement* of activities does change seasonally. Areas become more intensively used during certain seasons of the year, depending on the direction and intensity of the sun and wind, and the amount of precipitation—in other words, depending on the requirements for sheltered lifespace.

If some residences are only periodically occupied during the agricultural cycle, is it reasonable to assume that the site structure will reflect this

Figure 35. Dancing at a residence during a death fiesta. While men line up to enter a dwelling to drink corn beer (far right), a drummer and flutist play, and four men dance. The dancers are some of the soldiers in the Holy Week drama. The fiesta is one of three that will be held for the passing of the household's infant sons. Because the fiesta occurs during Lent, the dancing and music of the season become a part of the fiesta. The men dance in the Maintained Activity Area in front of the dwelling. The Swept Trash Area is at the edge of the corn field (squash is planted in this area in the spring). The Concentrated Refuse Area lies in the corn field. A path leads from the residence to the perennial stream where the household gets water. (Photograph by Jim Wolfgram)

intermittent occupation? While on the basis of a single year this may be possible, in an archaeological context it is unlikely for several reasons. First, while these residences may only be occupied for several days or weeks throughout the agricultural cycle, the "agricultural cycle" in fact extends over three-quarters of the year. Thus seasonal shifts of emphasis in activity space will occur at these residences as well as those occupied for longer time periods over a greater part of the year. Second, as stated above, whereas a residence may be occupied for the major part of a given year, this is not to say that it always has been or will be the main residence. Rather, the role of a given residence often changes over the years.

Winter residences tend to be both more season-specific and "topography-dependent." While not always the case, winter residences are generally abandoned during warm weather. Rather than being located in fields, winter residences are positioned where they receive early morning sun, and are close to a firewood supply and a water source. Activity areas at cold weather residences may be better defined because of the physical constraints of rockshelters, or because of other limitations imposed by the conditions given above.

Thus, the activity *structure* of winter residences is not significantly different from that at other kinds of residences. Differences between the three types of residences, if apparent, are more likely to be distinguishable in terms of the *seasonal* occupations rather than the *lengths* of occupation differentiating the main residence from one occupied during the agricultural cycle. As archaeologists, we need to be able to distinguish site structural patterns associated with the physical constraints of rockshelters from information relating to season of occupation. In other parts of the Sierra, Rarámuri have used rockshelters as residential space during warm weather over and above occupation during religious celebrations (e.g., Lumholtz 1894, 1902). We can not assume rockshelters are only used as cold weather residences.

What does the variability in Rarámuri residential mobility strategies mean from the perspective of archaeological sites? Given the nature of activities occurring at these agriculturalists' residences and the variety of ways in which each might be used during the history of its occupation, it may be quite difficult to distinguish summer from winter residences, or the main residence from one occupied periodically during the agricultural cycle. Detailed data collection of household assemblages and the distribution of artifacts at residences might prove this characterization to be overly pessimistic. A major goal of this study, however, is to demonstrate that at a more general level the distribution of materials at a Rarámuri residence *can* be informative about settlement organization. We can then take our understanding of this information back to the distribution of the materials we find in archaeological contexts.

The organization and use of space at a residence is in part dependent on factors of both subsistence and settlement organization. The activities occurring at a residence, the physical accommodations for those activities and the character of the group living at that place influence residential site structure. The high degree of residential mobility among the Rarámuri is manifested in the distributions of their furnishings and equipment, here specifically in their organization and use of residential space.

Rarámuri Residential Site Structure

The composition of residences refers to the elements of site structure typically found at the residence. The following elements compose Rarámuri residences: dwelling structures, sheltered workspace, residential activity space, storage facilities, other structures, trash disposal areas, and gardens and fields. Figure 36 shows the spatial components of the Rarámuri residence. The rest of this chapter discusses these components, and where activities at the residence are carried out.

The Rarámuri Structural Activity Area

The Structural Activity Area consists of the residential dwelling(s) and the area immediately surrounding it. Here, Rarámuri perform the majority of activities occurring at the residence, including food processing and preparation, recreation and social interactions, sleeping, and craft manufacturing tasks. They carry out activities both within and outside of the dwelling. Sheltered workspaces are often close to the dwellings, and are included in the Structural Activity Area.

Dwelling Structures

Dwelling structures are the major form of sheltered lifespace at Rarámuri residences. These structures house sleeping facilities, food processing and preparation facilities that include both grinding and cooking facilities, and numerous stored goods and foods. Moreover, much of the "living" (inter-

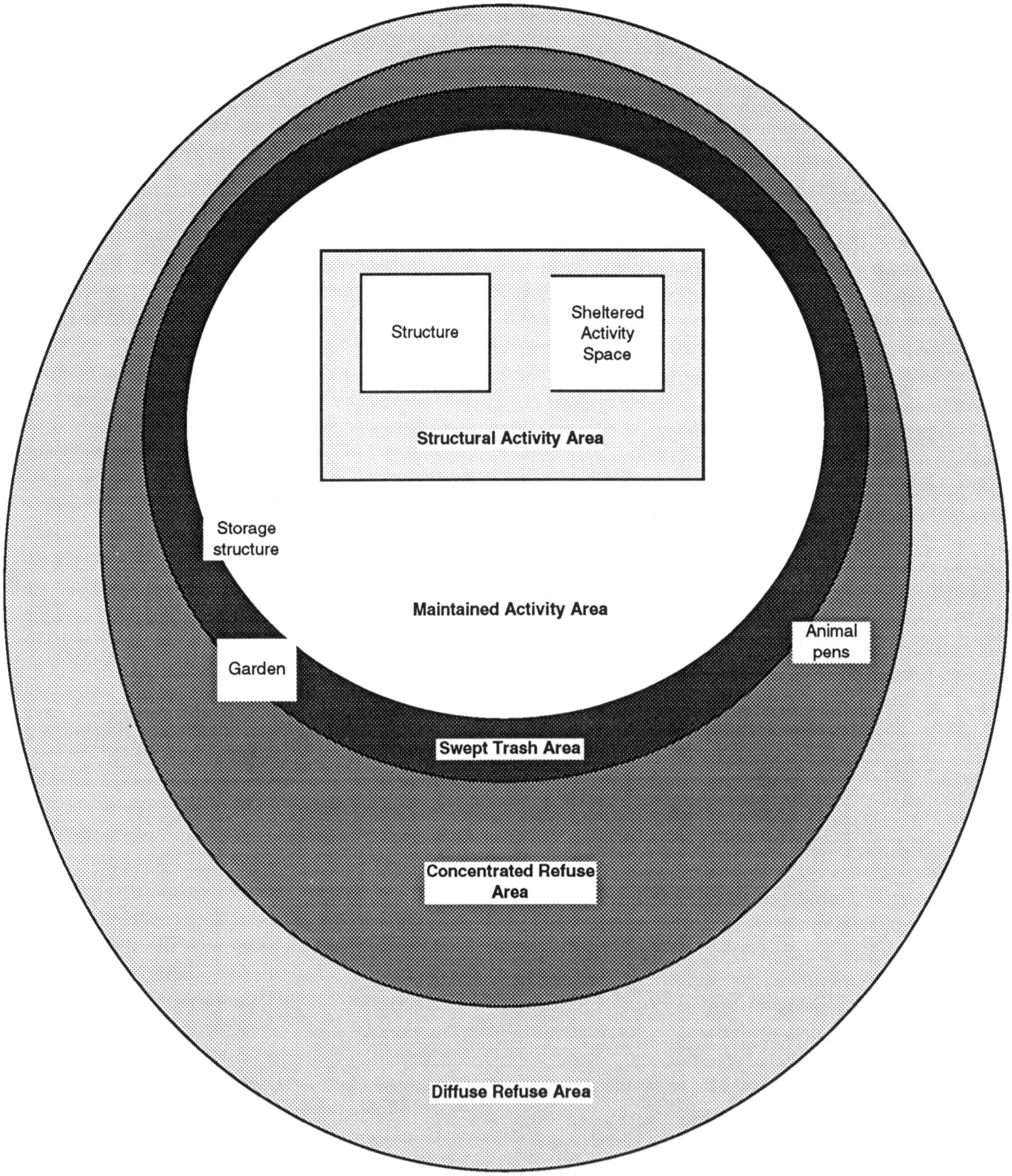

Figure 36. Spatial components of the Rarámuri residence.

acting, visiting, socializing, and child caretaking) occurs here. Dwellings provide shelter from the elements for householders and protected storage space for various types of foodstuffs (Hunter-Anderson 1977).

Another important function of the dwelling is to shield people and activities from the eyes of other community members. One principal means of achieving privacy is to go inside the dwelling and shut the door. This is one way of signifying that one "isn't home" for company. In Rejogochi, eating special treats frequently took place inside, and small animals are processed indoors (e.g., plucking a chicken), to keep others from knowing about it.

Rarámuri houses in the study area are commonly made of logs, although some are made of, or incorporate, stone in the structure. A very few houses are of adobe brick. The houses are roofed over with hand-hewn planks of wood bevelled toward the center. The planks overlap each other, with the bevelled side alternating facing up and down. Ideally, this arrangement results in a leakproof roof. When weather permits, Rarámuri will sometimes remove one or more planks from the roof, or flip them back on top of the others. This opening, or "skylight," provides better light and ventilation to the interior of the dwelling. These openings are most generally above the hearth, allowing not only for better lighted workspace, but also a more efficient exit for the smoke.

The traditional means of heating a dwelling is an unprepared hearth at the level of the ground surface (Figure 37). This hearth is also the primary cooking facility. The hearth is either near the center of the dwelling or against one wall. When the hearth is in the center, and the roof is intact, smoke may exit through an opening left in the offset between the two halves of the roof (one side overlapping the other). The smoke exits not only through the roof, but also through the unchinked walls of the structure. In a number of cases, the dwelling with the hearth is either not completely roofed or not completely walled (e.g., Structure B of Residence 5, Figure 26). Some dwellings with the hearth against the wall will have a crude chimney above the fire.

Figure 37. Hearth inside Rarámuri dwelling.

In windy, wet weather, these unchinked dwellings are uncomfortable and drafty (one reason for moving to rockshelters in the winter). At the same time, however, they offer a fundamental service to their occupants: providing many little windows on the world. The Rarámuri are adept at keeping track of the comings and goings of people and animals through these crevices, and even tightly constructed houses will have strategic areas left unchinked for this reason.

Dwellings with stoves are far more comfortable in inclement weather. I observed more variation in the placement of these facilities than of open hearths. I noted stoves in the center of the house, in one quadrant, and along a wall of the dwelling (Residence 16, Figure 32, and Residence 5 Structure A, Figure 26). Almost without exception, the stove has a pipe that is more or less effective in funnelling smoke up and out of the dwelling. The stove generates more heat than do hearths, but also consumes more wood. These dwellings are more tightly constructed than those with open hearths. Their walls are closely fitted with well-chinked boards or logs. They tend to be warm and dark. Stoves are set on the ground, or (rarely) on a low platform, and easily moved either to another structure, or simply out of the way, if necessary.

Stoves are becoming more common in Rejogochi. Frequently, when a residence has two dwellings, one has a stove, and the other has a more traditional hearth. Having the two types of facilities allows for both function- and season-specific uses of the dwellings. At such residences, the dwellings with stoves are often used as sleeping places and, more frequently than in other seasons, cooking places in the winter because of the increased warmth. The other dwelling is the main activity place for food processing, some preparation and cooking, visiting, leisure, and activities requiring light. Residence 5 (Figures 13 and 26) was used in this way, with food processing and preparation, and other activities occurring more frequently in the Maintained Activity Area or in Structure B than in Structure A. At different times of the year, the functions of the paired dwellings change. For example, during the summer, Structure A at Residence 8 (Figure 33) was closed and used almost exclusively as a storage area.

Many residences have a single dwelling (e.g., Residence 12, Figure 19), and residences with two dwellings do not always have the same seasonal/functional variability described here. Households with residences having two structures do not always occupy them year-round rather than moving to a cold weather residence in the winter (e.g., Household 2, with Residences 5 and 4, Figures 13 and 28). During my fieldwork, the majority of people with two dwellings at their residence also moved seasonally (Tables 1 and 2; see also the discussions above of Households 3 and 11).

Multiple dwellings (i.e., structures) at residences are considered the functional equivalent of multiple rooms. But although a specific dwelling may have features making it conducive to certain activities during parts of the year, no single structure can be identified as consistently and/or exclusively season- or activity-specific. In March 1983, the structure with the stove at Residence 16 was used for almost all indoor activities (Figure 21). The second room was used for storage of clothing, food, cookware, and for grinding with the molino. During other months of the year, however, the household used the interior room for sleeping space (W. Merrill, personal communication, 1989). The small window allowed additional light into the dwelling.

Dwellings are also used for bulk storage. This is the case with Residence 18 (Figure 23). Structure C, the smallest and oldest dwelling at the residence, is no longer used as dwelling space. The building contains the molino used by the household, and in the fall the dwelling is filled with squash, apples, clothing, ceramics, and various tools. The use of old dwellings for storage was also observed at Residence 19 (Figure 38). Household 11, a childless married couple, used Structure A as their dwelling, and Structures B and C for storage. Structure B also housed the woman's molino. As the center of domestic activity for Household 6 changes from Residence 12 (Figure 19) to Residence 14, the former is used less as a dwelling and more as a large storage unit, and the household spends increasingly more of its time at the daughter Irene's (Household 7, Figure 18) newer Residence 14. Residence 14 lies approximately 75 m from Residence 12, with a field separating the two houses. During 1987, harvested squash and a number of boxes of apples were stored in the older dwelling while Household 6 continued living at Residence 14.

In an ethnoarchaeological study of maize storage strategies among the Maya, Gotthilf (1982) discusses the heat damage to corn. A warm environment encourages bacterial activity and insect infestation; it may dehydrate the kernels below the level for successful germination. Among the Maya of her study area, these were important considerations for not storing maize near cooking areas on a long-term basis. If the Rarámuri are actively and frequently using a place, in this case a dwelling, for cooking and/or heating purposes, then that structure's utility as a storage facility is expected to be minimized. The Rarámuri dwellings mentioned as bulk storage facilities are all dark, cool, and fairly well protected against the elements.

Beyond bulk storage units at residences, small and miscellaneous items are stored in the nooks and crannies of dwelling walls, under the low beds, between the roofs and ceilings of dwellings (if the dwelling has a ceiling), and on the roofs of structures. Some dwellings have shelves, either along a wall or across the corner of a structure (e.g., Figures 26 and 32). The former type tends to be for storage of small and miscellaneous items (for example, salt, cups, jars, and batteries), while the latter is frequently a station at which the molino is set up, as well as a storage place for miscellaneous items and some foodstuffs.

Sheltered Workspace

In addition to dwellings, Rarámuri residences have sheltered activity areas usually consisting of one or more walls built as windbreaks but often without cover overhead. The windbreaks are commonly rock and/or branches. As in the case of Residence 8 (Figure 33), the sheltered workspace often incorporates the wall of a dwelling into its structure. In other cases, the workspace is discrete from

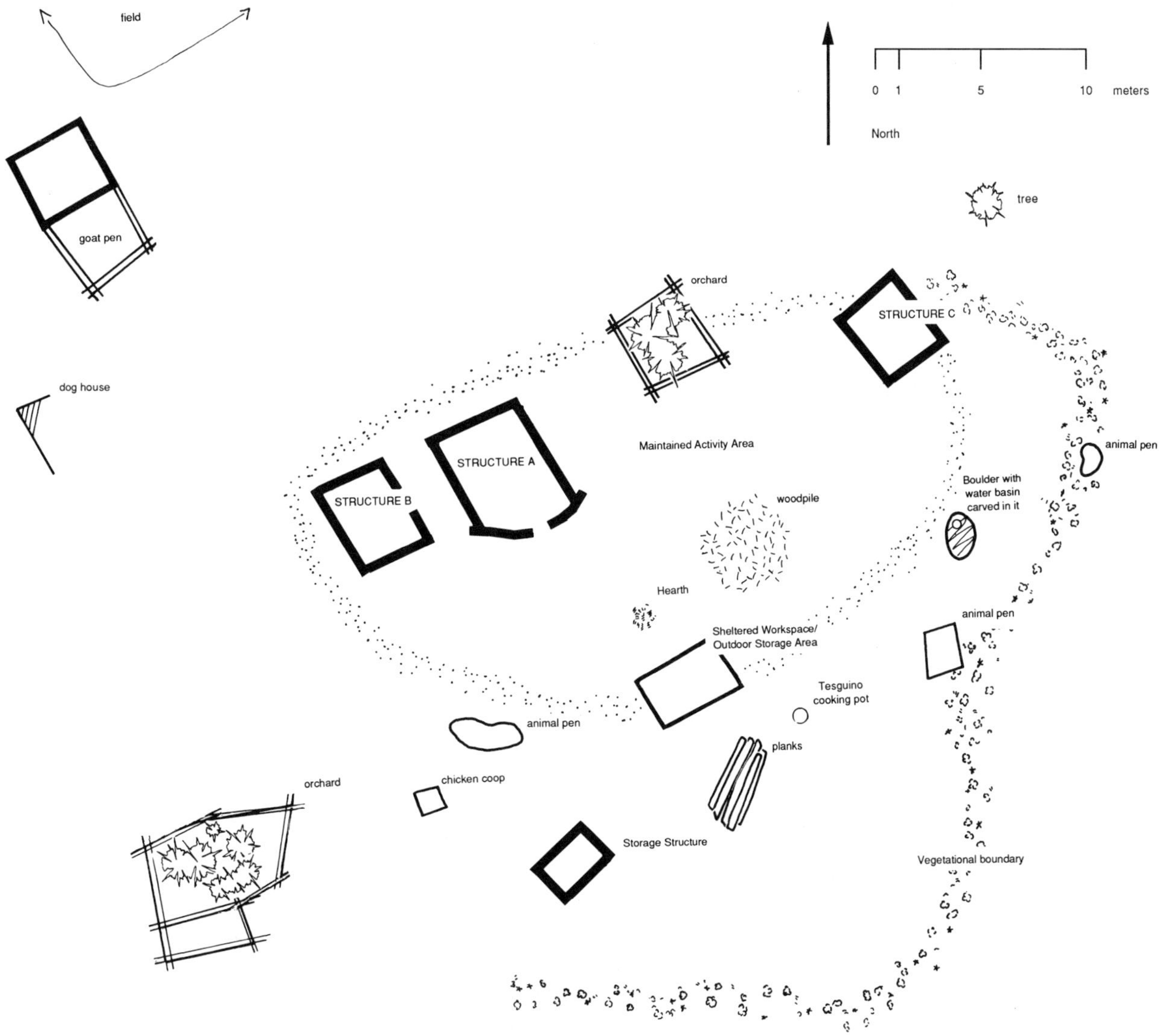

Figure 38. Residence 19.

the dwelling and the Maintained Activity Area (e.g., Residence 17, Figure 34).

Some of the most common activities occurring at these work places are weaving and ceramic manufacture. These areas provide protected space where one can set up a loom, or set out the items necessary in making ceramics. The walls of this sheltered space act as barriers to the elements, and to a limited extent, to people and animals; they re-route traffic around this workspace.

These sheltered workspaces are often "permanent." With any element of Rarámuri residences permanence is qualified because everything at the residence, including dwellings, are subject to removal, modification, and realignment. With the exception of the most ephemeral of windbreaks, modification of the physical structure or arrangement of these work places is uncommon.

The use or function of these activity areas is quite flexible. In some cases the workspace is abandoned for a time and the activities performed elsewhere. For example, from January to June 1987, Santa Rosa's loom was in a sheltered activity area (Residence 8, Figure 33), protected to the east by the wall of Structure A and to the north and west by a single course wall of boulders (ca. 50 cm or less high). These walls may have been more substantial in previous years, but this is not certain. When this activity area was first recorded, planks propped against the roof of Structure A to the northwest corner of the rock alignment provided additional shelter. An old bed from the mission school was put ("stored") here in June 1987. Santa Rosa moved the loom southeast, to the south side and southeast corner of Structure B. In the fall, the loom was again moved, this time to the east side of Structure A. In both of these subsequent locations, boards, corrugated tin, blankets, small trees, plastic, and other materials were used to fashion temporary shelter in the area around the loom. The bedstead remained in the low-walled activity area, and no activities were subsequently performed there.

At the time I was mapping Residence 18 of Household 10, a loom was in use in the "walled activity area" between Structures B and C (Figure 23). The loom was later removed, and never put back in this workspace during the course of my fieldwork. During parts of the year, boards used as a bed were laid in this area. This section of the workspace was also used to store sprouting corn kernels for making corn beer. The walled activity area was also the locus of ceramic production. Katilachi used this area for many food processing and preparation activities during much of the year.

Residences 12 and 18 (Figures 19 and 23) both have sheltered work space containing a hearth close to the dwelling. A number of food processing and preparation activities, especially cooking, occur here during the year. To a lesser degree, these same places are also used as sleeping space. In other words, when the weather is suitable, some of these "sheltered workspaces" serve the same purposes as the dwelling. For example, a second outdoor workspace was established east of Structure A, Residence 18 (Figure 23) for most of the summer and fall. A narrow bed was set against the east side of Structure A, built up on its southern end and roofed over, with a hearth east of the shelter. This area, although not entirely constrained or defined by structural elements, was used as sleeping, food preparation, and leisure space by members of Household 10.

During the height of the rainy season, these activity areas are not used for cooking or sleeping. While sheltered workspaces incorporate various materials as protecting walls, the Rarámuri of Rejogochi rarely cover these areas with substantial roofs. Unprotected hearths are often inundated with water and the wet ground makes it difficult to produce a good cooking (or warming) fire. This is different from the ethnographically documented use of outdoor hearths in the American Southwest, which has been used by archaeologists to infer summer occupancy at archaeological sites (Powell 1983). Variations in precipitation and in sheltering techniques are partial explanations for the differences in the uses of outdoor hearths in the summer. In Rejogochi, the Rarámuri use sheltered workspaces with outdoor hearths for cooking and sleeping, most extensively in the fall, when the rainy season ends, and to a lesser extent in the spring, when the weather warms. The Rarámuri may use these areas for other activities regardless of the season, however.

The Rarámuri Maintained Activity Area

The Rarámuri carry out many activities at the edges of the open space around structures that, with the buildings themselves, make up the Structural Activity Area. The structures are used for shade, as backrests, and protection against the elements. The clean, bare area encompassing the dwelling and sheltered workspace is referred to here as the Maintained Activity Area (Figures 9, 13, 19, 21, 23, 28-31, 33-35, 38). Activities concentrate near the Structural Activity Area and grade out into the Maintained Activity Area. Many activities shift seasonally between the Structural Activity Area and the Maintained Activity Area.

The Maintained Activity Area is probably swept at least once or twice every couple of weeks, though degree of cleanliness ranges widely among individual Rarámuri. Maintenance also includes putting away materials after an activity is complete, or at the end of the day. The Rarámuri maintain this area more frequently than elsewhere on the residence, including the interiors of the dwelling or sheltered workspace. In addition to this maintenance, the area is regularly and thoroughly scouted by animals scavenging for food, who peck, eat, root, and lick the activity area clean.

Dogs and pigs play an important role in keeping residences free of human waste. Adults of a household use a place some distance from their residence for defecation, but children remain close to the residence. "Potty-training" young children generally consists of teaching the child to go beyond the boundary of the Maintained Activity Area.

Because animals wander freely through the maintained activity space, food items and containers are not left open or unattended in these areas. Human and animal traffic in and around this area tramples artifacts and foodstuff into the ground. As a result, the artifact assemblage that remains in this area is of relatively smaller size than in other parts of the residence (Binford 1987; Dodd 1984; O'Connell 1987; Schiffer 1987).

When large numbers of people are at the residence, as is the case when preparing for and hosting work parties or religious celebrations, the Maintained Activity Area serves to contain the additional people and the overflow of activities more commonly performed in the Structural Activity Area (Figures 29 and 30).

Visitors perform household tasks both indoors and out, for example, processing corn and wool. Sexes tend to sit in separate groups at these times (Figure 39, see also Figure 35). The Maintained Activity Area is an ideal place for dances and other religious activities during the growing season. After harvest these events will also take place in the fields surrounding the residence (see Religious Activities below).

Storage Facilities and Other Structures

Close to the interface between the Maintained Activity Area and the Swept Trash Area (see below) are arrangements of various outbuildings. These include the storage structure, the corn drying bin, and the chicken coop.

Corn is stored on the cob in bulk storage facilities. In Rejogochi, these are discrete structures made of wooden boards, notched at the ends and close fitting (see Pennington 1963:58-61 for a discussion of Rarámuri storage structures). A small doorway with a trick lock or a small padlock is the usual means of entrance and exit. Items that are somewhat bulky or used with some frequency but not on a daily basis are stored in the area between the top of the ceiling and the roof. Examples of these include unused ceramics, tools, and craft items. Rarámuri sometimes take advantage of natural rock formations, constructing a protected (and camouflaged) storage niche by closing the opening with rock and mortar. Few of these storage facilities are currently in use in Rejogochi, but there are a number of them abandoned in the valley.

Most generally in Rejogochi today, the storage structure is located at the residence. Sometimes it is in a nearby field. The Rarámuri of Rejogochi transport corn from the field to the storage unit on their backs. Only occasionally do they use horses or burros as draft animals, or wheelbarrows. A household has several options regarding where to locate its storage facility. If the household harvests the corn and then transports it to the storage structure at the residence, the transferral must be done at harvest time. If a household harvests the corn and leaves it at or near the field, it is required to go down periodically to resupply the stores at the residence. If the storage structure is at the field, the household also runs the risk of having their corn stolen (the fate of one household in Rejogochi when they left the community for several weeks).

Residences used during the agricultural cycle also include storage structures. Neighbors and relatives watch the structure when the household is not in residence. The family uses this harvest while living in the community, tending their field or visiting. The household may carry corn from this storage structure back to its main or cold weather residence, if necessary. Alternatively, the household will move to this residence when it has run out of food at another one, rather than transport the corn from one place to the other (Hard and Merrill 1992).

Various other structures at Rarámuri residences are most often at the periphery of the sites. These include structural elements used to create ephemeral or temporary sheltered activity space in the Maintained Activity Area (e.g., Figures 23 and 28), and corn drying structures and animal pens (e.g., Figures 19, 21, 25, and 34). Areas designated as garden plots are often surrounded by structural elements (fences and walls) to protect seedlings and saplings from hungry animals (e.g., Figures 31, 33, and 38).

"Ephemeral (or temporary) structure" is perhaps too strong a term to describe the conjoining of structural elements into shade and shelter facilities. These elements include miscellaneous boards, pieces of corrugated tin, rocks, blankets, and posts put together to form sheltered work and sleeping space within the Maintained Activity Area. For example, during much of the summer and fall of 1987, part of Household 10 used an outdoor hearth at Residence 18 in front of Structure A (Figure 23). Not only were boards laid out on the ground both for sitting and as a work surface, but the household installed an elevated bed, roofed over and shaded from the direction of strongest sunlight. In the winter, although the hearth continued to be used on occasion, the shelter itself was dismantled. Also at Residence 18, boards leaning against the storage structure create additional sheltered storage space—for clothing items, for miscellaneous tools, and also to keep firewood and cattle dung dry (the latter is stockpiled for use in firing ceramics).

Ephemeral shades and shelters usually incorporate part of a structure as a wall and/or support for the rest of the elements of the shelter. The elements, and the shelters themselves, are moved and reconstructed frequently, leaving little evidence of their existence after the structure is dismantled.

Outdoor corn-drying structures are large bins where the corn is put before being put into the

Figure 39. Women gathered at a drinking party. The women sit near the edge of the Maintained Activity Area and Swept Trash Area. The windbreak to the right of the dwelling was constructed in this section of the Maintained Activity Area when the women moved their metates into this area. Here, they ground and cooked the sprouted corn to make beer. A covered jar of corn beer sits next to the structure. The roof of the dwelling is used to store foods (in the bucket and basket) and personal gear of both the household and its visitors.

storage structure. These structures may be constructed at harvest time and later dismantled, or left standing year-round. Some drying structures are elevated, while others sit relatively close to the ground. Usually these facilities are frequently placed near the storage structure. They may be close to the boundary of the Maintained Activity Area, or somewhat removed from the residence. In some cases, a cone-like structure of poles is set up at the fields where the maize is allowed to dry before it is transported to the residence for storage (see Pennington 1963:59).

The Rarámuri Swept Trash Area

Circling the Maintained Activity Area is a high concentration trash zone created through the sweeping of debris out of the Maintained Activity Area (Figures 13, 39). This band of trash denotes the edge of the residential space, and beyond it lies unmaintained space; only a few activities occur beyond this border. Chopping wood (although the woodpile is also located in other areas), animal processing, children's play, dancing and other religious events, and defecation are some of the activities performed beyond this boundary.

Traditionally, it is impolite to approach a residence without an invitation; it is my impression that the Maintained Activity Area generally constitutes the edge of the residence, or "home." The Swept Trash Area acts as a social threshold, requiring permission to cross.

The Rarámuri Concentrated Refuse Area

A more generalized trash area lies farther out from activity areas, beyond the band of swept trash (for example, Figures 25, 31, 33, and 34). While the Swept Trash Area is predominantly formed by sweeping the Maintained Activity Area, the Concentrated Refuse Area is the area surrounding the residence that contains the majority of the trash at the site. This area is created through the placement and movement of trash out away from the Maintained Activity Area. Many small items associated with activities in activity areas fall in the maintained space, and during the course of maintenance are swept out into the trash zone ringing the maintained work space. Larger items are more frequently picked up and placed, or tossed, farther away from the activity area. Due to traffic in and around the residence, the play activities of children, and the basic orientation of activities at the residences, trash accumulates in certain areas, some accruing greater concentrations of debris than others (Figure 23). The greatest density and areal extent of debris lies in front of that part of the residence with the highest frequency of activities (e.g., Figures 21 and 31).

High density pockets of debris occur within the Concentrated Refuse Area, but these are more likely to be the result of topographic variation or the relationship of the area to a favorite work place than of discrete dumping episodes. For example, at Residence 8 (Figure 33), the trash is most dense in the area beyond the doorway of the dwelling and Maintained Activity Areas. A slope down from several areas of rock outcrop establishes a natural break between activity area and trash area, while the slope and its base provide a direction and backstop for the trash as it "flows" from the site.

In contrast, the large trash area at Residence 18 (Figure 23) is also out in front of the house, but the slope of the land is not such that trash is readily directed outward. Consequently, at this residence the distribution of trash is greater and more diffuse, with an area of medium and medium-to-light trash scattered about. This area catches the majority of trash simply because it is located in front of the majority of activities. The medium and medium-to-light trash distribution is created through the scattering of trash by traffic going in and out of the center of the residence, by maintenance activities, and by the same kind of trash flow from the Maintained Activity Area in parts of the residence where there is a lower intensity of activity space use.

The main concentration and other trash scatters of Residence 11 (Figure 31) are smaller than those of either Residences 8 or 17 because Residence 11 is closer to the edge of the perennial stream running through the valley. In front of Residences 11 and 18 the stream is entrenched; a short (approximately 2.5 m vertical) climb is necessary to get from the stream bank to the residence. Much of the trash accumulating close to the edge of the entrenched stream bank falls over and is washed away eventually.

The Concentrated Refuse Area represents high diversity in both the associated disposal behaviors and the nature of its contents. Trash from the Swept Trash Area grades into the Concentrated Refuse Area through human and animal traffic, children's play, common residential activities, and also by natural means. Larger trash items are tossed or placed directly in this area. Potentially useful items (for example, cans, bottles, and large pieces of hardware) are discarded in this area in much the same way as other trash but are also subject to retrieval when needed. This behavior, called "provisional discard" (Deal 1985; Hayden and Cannon 1983; Killion 1987) or "passive storage" (Binford 1977, 1979; Brooks and Yellen 1987; Chang 1988), means the Concentrated Refuse Area acts as a storage area for items to be retrieved and drafted into use at a later date.

Animal pens tend to be at a distance from the residence, although chicken coops may be relatively close, providing greater protection from coyotes (e.g., Figures 23, 33, 34, and 38). Indeed, at Residence 11 (Figure 31), the chickens roosted most often on the shelf between the two structures. If a hen is laying, or new chicks are about, a more protected place than the chicken coop might be devised. Under certain conditions, the Rarámuri will make exceptions to the placement of animal pens. For several weeks during spring of 1987, Household 5 used one of its dwellings as a barn. The household built a pen in the southwest corner of Structure B, with the boards that ordinarily served as bed and bench facilities. This dwelling was used, I was told, because there was no other warm shelter for the animals during the cold nights. Earlier that winter, the household had used Residence 9, a rockshelter, that has a corral for herds of sheep and goats. The household offered no explanation as to why the rockshelter was not reoccupied at this later time. The pen inside Structure B at Residence 11 was then moved outside. The household incorporated the southwest wall of Structure B as part of the pen. It was at this time the residence was mapped (Figure 31). The loom, originally in the space occupied by the pen, was moved to the southwest wall of Structure A. After a couple of weeks, the household dismantled the pen and the animals were corralled on fallow portions of the household's field. The household removed the manure from the area of the pen, spreading it in a nearby portion of their field.

Pig pens are set at a distance from the dwellings and general Maintained Activity Area of residences. If the residence is close to the perennial stream or another sharp drop-off, the pens are often placed near the edge—presumably because this allows the debris in the pen to fall away, with minimum caretaking of the facility (Figure 23).

The Rarámuri Diffuse Refuse Area

The Diffuse Refuse Area is the relatively light trash scatter encircling the residence, generated in large part through the dispersal of trash out from the Concentrated Refuse Area (for example, Figure 23). The major causes of this dispersal are disposal of debris by tossing, the shifting of trash by the human and animal traffic through the residence, and natural movement downslope. Plowing and garden cultivation at the margins of the site contribute to the spread of trash from the Concentrated Refuse Area and, to a more limited extent, the Swept Trash Area, into the Diffuse Refuse Area. During the growing season, plants in the garden and fields act as barriers, trapping trash and limiting its dispersal.

Gardens and Fields

Many households in Rejogochi have a garden plot associated with their residence. These gardens contain squash, some corn, occasionally condiment plants (for example, chili, garlic, and tomatoes), and vegetables from seeds given by the mission. Areas with concentrations of organic trash in the soil are used as gardens in the spring. Residence 8 was mapped in April of 1987 prior to the planting of any garden (Figure 33). By June of that year, Rafael and Santa Rosa had planted squash in the area within the Maintained Activity Area identified as the woodpile, and in some of the area around it. They established a second woodpile to the south of the Maintained Activity Area, in the general trash area. Later, they also planted a section of this general trash area. Santa Rosa and Rafael used the area of high density trash to the east of the Maintained Activity Area, on the slope below the rock outcrop, growing corn and squash. An old animal pen protected special plants—a bulb of garlic, one or more fruit tree saplings, and some tomato plants. At Residence 18 (Figure 23) the medium/light trash scatter north of the fence was planted as a garden, predominantly of squash.

Many households have similar gardens planted in protected, enclosed areas, and also in the rich organic soil around the periphery of the residence. Typically, Rarámuri in Rejogochi plant small bean plots and squash gardens at the margins between the field and the residence. When maize itself extends into this margin of the Maintained Activity Area, there is a noticeable difference in the height of the plants.

The majority of residences currently occupied in Rejogochi border a maize field (e.g., Figures 13, 19, 23, 31, 35, and 39). This has not always been the case; a number of abandoned sites in the valley are at some distance from arable land, towards the heads of small drainages, or up on slight knolls above the valley bottom. Possibly these sites were occupied when the population of the valley was greater than it is now, before a flu epidemic in the early 1900s substantially lowered the population (W. Merrill, personal communication, 1989).

The current practice of having one's residence in or beside a field has some consequences relevant to a discussion about space use. The actual size of the residence, that is the size of the Maintained Activity Area and general trash catchment area, tends to shrink during the growing season (Figure 40). Residence 18 (Figure 23) was mapped in February 1987. By summer of that same year, all the medium/light trash scatter north of the fence was planted as a garden of squash and vegetables. Squash was also planted bordering the Maintained Activity Area to the southeast of Structure A. The field to the east of the residence extended into what, during the non-growing season, is maintained activity space. Space classified as "non-maintained space" during the early part of the year was used as activity space during the growing season; socializing and weaving both took place under the fruit trees located at the residence, but outside the "Maintained Activity Area."

Seasonal Aspects of Residential Elements

With increased vegetation around the residence during the growing season, there is also increased activity in these areas. Small-scale harvesting of garden crops is a more or less daily event, as food for meals (and snacking) is picked and processed. Defecation in proximity to the residence increases due to the cover provided by the maize plants. Fields become a place where Rarámuri discretely meet to have sex, and the vegetation surrounding the residence becomes a playground for children, who play

Figure 40. The edge of the Maintained Activity Area at the height of the growing season. The corn and squash in the field adjacent to this residence have been planted in the Concentrated and Diffuse Refuse Areas. The boards at the bottom of the photograph have been placed in the Maintained Activity Area as an outdoor bed and work surface (the corn plants will provide shade here during parts of the day).

hide and seek and other games here. Food processing activities move to the interface between the Maintained Activity Area and the field in order to take advantage of the shade cast by corn plants.

In considering seasonal variability in residential space use and the influence of the gardens and fields on Maintained Activity Areas, two points should be kept in mind. First, the amount of *activity space* available changes during the year—perhaps more than does the *activity structure* at the residence. Second, activities occurring at the changing boundaries of the spatial components will further modify the distribution of materials at the residence. For example, plowing up to the edge of the Maintained Activity Area blurs the Swept Activity Area and spreads refuse from the Concentrated Refuse Area even further into the Diffuse Refuse Area.

Chapter 4

A Comparison of Space Use in Three Agricultural Systems

The extreme flexibility of space use at the Rarámuri residence, as well as the range of variation possible in Rarámuri mobility strategies, was well demonstrated in the preceding discussion. Flexibility and variation exist in all societies, but the *degree* of flexibility is associated with other aspects of the system. This relationship is evident in a comparison between the Rarámuri and two other agriculturalist groups, the Guarijío and the Tuxtleco. Both groups are subsistence or semi-subsistence agriculturalists, primarily dependent upon maize. In both cases, they are more residentially sedentary than the Rarámuri.

While there are a number of similarities in space use among these agriculturalists, there are also important differences. Residential mobility, intensification of agriculture, population density, climate, and topography are all factors influencing residential site structure. By comparing the residential site structure of more sedentary agriculturalists with Rarámuri residential site structure, not only can we identify how these factors are expressed through the organization and use of space, but we can also develop criteria for distinguishing among various subsistence-settlement organizations.

I chose the Guarijío and Tuxtleco for comparative purposes to keep several critical factors constant. Specifically, the Rarámuri, Guarijío, and Tuxtleco all remain relatively independent of Mexican economy. All three groups subsist on corn, supplemented by other cultigens, some wild foods, and a small amount of meat from domestic animals. Moreover, food processing and preparation methods are similar among the three groups. Food processing and preparation are major activities at residential sites, and together a primary influence on space use. Keeping these aspects constant for the three groups allows us to separate out the role of other factors, including mobility, in influencing residential site structure.

The Guarijío

The Guarijío are a small group of sedentary agriculturalists living on the western slopes of the Sierra Madres, Chihuahua, Mexico, in the upper and middle valleys of the Rio Mayo (Figure 41). Known variously as the Warihio, Varohio, Huarahia, and Guarihia, very little anthropological interest has focused on the Guarijío. Until recently, investigations of the Guarijío centered primarily on their language, a Uto-Aztecan derivative, and its relationship to the languages of more well known neighbors—e.g., the Rarámuri. Howard Gentry's 1963 monograph, based on fieldwork done between 1934 and 1936, is the only published ethnography devoted to the Guarijío to date. Ethnoarchaeological research by Walter Dodd from May 1981 through June 1982 is available in a number of publications (Dodd 1984, 1985, 1987, 1992). The current discussion relies heavily on Dodd's work in the Guarijío communities of Chiltepín and La Barranca (Dodd 1984, 1985, 1987, 1992, and personal communications 1985-1989), supplemented by Gentry's ethnography (1963).

Roughly 2000 Guarijío occupy the western slopes of the Sierra Madres in a diverse and rugged terrain. The vegetation is both oak woodland and tropical deciduous forest (Dodd 1984; Gentry 1963). Temperatures range from 30 to 100 degrees Fahrenheit (approximately -1 to 38 degrees centigrade), averaging between 60-90 degrees Fahrenheit (16 to 32 degrees centigrade) during most of the year. Guarijío terrain is generally extreme, with both fields and communities occupying steep slopes.

The Guarijío live year-round in small dispersed communities, producing corn, beans, and squash for their own consumption. They supplement this diet with wild plant and animal foods, and domestic animals. Besides cattle, goats, pigs, and chickens, the Guarijío keep dogs and cats. The Guarijío of La Barranca and Chiltepín incorporate a variety of wage-earning strategies into their subsistence regime, some households underpro-ducing agricultural produce and supplementing their subsistence with wages or by selling goat milk and cheese. Other households deliberately overproduce and sell the surplus to their neighbors (W. Dodd, personal communication, 1989).

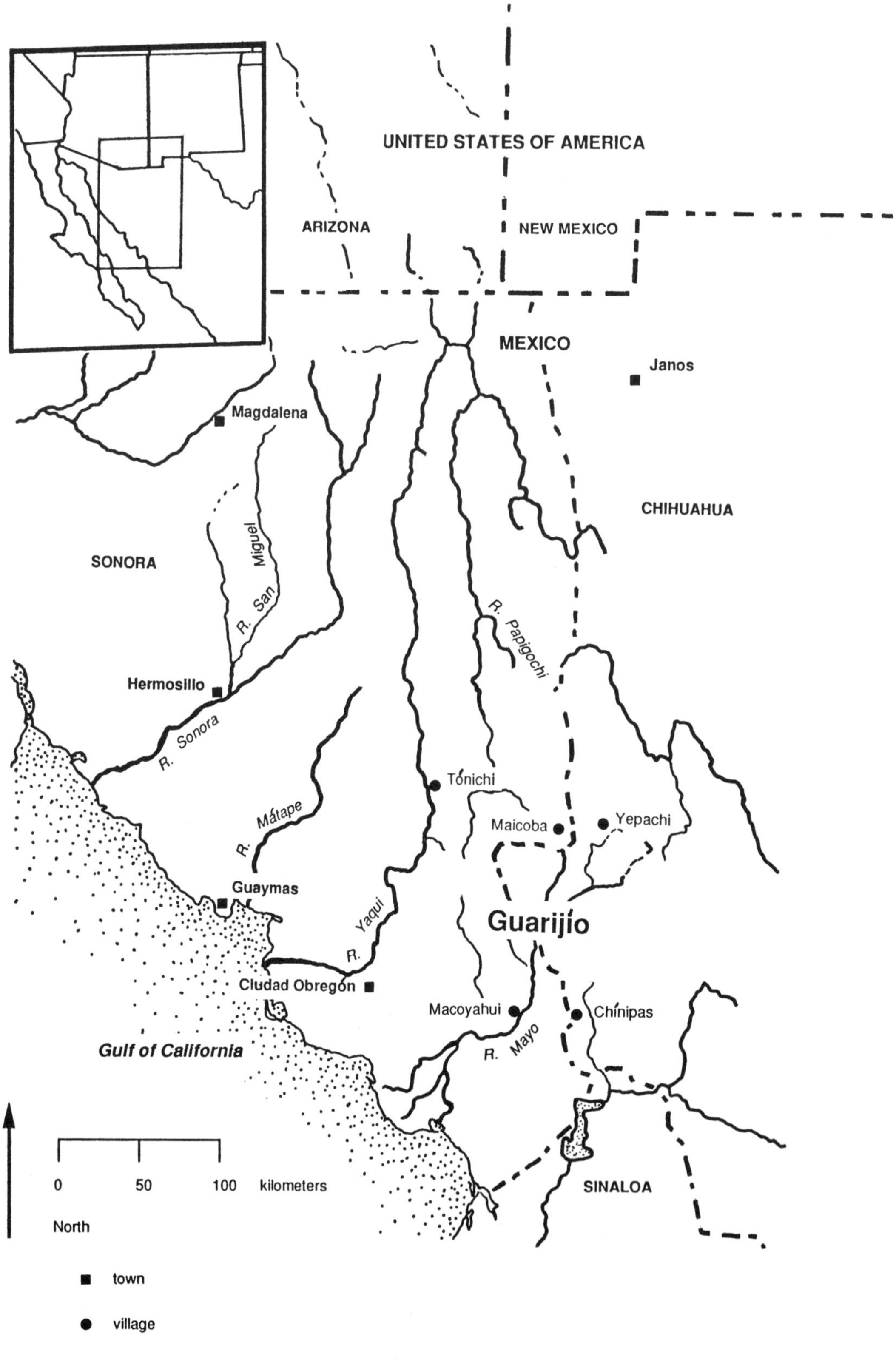

Figure 41. Location of the Guarijío homeland (in part, after Hinton 1983:Figure 1).

Guarijío residences are generally excavated into the steep slopes of the mountains (Hinton 1983:Figure 7, p. 326). Clusters of households concentrate at a convenient distance from a permanent water source. Water resources are relatively scarce. Houses are located in proximity to springs, seeps, and streams, although fields are not. Dodd has suggested that the distribution of permanent water resources in Guarijío territory is an important factor in influencing the distribution of settlements across the landscape (W. Dodd, personal communication, 1988). The combination of topography and available water are important reasons for the arrangement of Guarijío communities, which are more densely populated than is Rejogochi and other Rarámuri communities.

At Chiltepín and La Barranca, fields are usually within 10-25 minutes' walk (ca. 1.6 km) of the residence, but can be as far away as 3 km. Fields are tended on a regular basis; workers make trips out to the field during the day, and return to their residences in the afternoon or evening rather than spending the night at the field. Fieldhouses and natural shelters used as fieldhouses are virtually unknown among the Guarijío although temporary shelters are used in other contexts (W. Dodd, personal communications, 1988 and 1989).

The majority of Guarijío households in La Barranca and Chiltepín are far more sedentary than are the Rarámuri of Rejogochi. Dodd has estimated the fully sedentary Guarijío at 50 percent or more (W. Dodd, personal communication, 1985), and adds:

> If I count as a residential move all situations where a homestead is abandoned and subsistence is thereafter performed or gained elsewhere on a regular basis (not including short visitations or fiestas), the mean number of moves per residential group per annum for the Guarijío I observed is probably much less than 1. (W. Dodd, personal communication, 1988)

A few households in La Barranca and Chiltepín do move during the year. One reason given for making seasonal residential moves is to take advantage of greater solar exposure; these moves (of 500 m or less) are to other habitations within the community. The Guarijío also make seasonal (winter and spring) moves to the Mexican town of Cinerea or elsewhere for wage labor. Feuding and discord within the community, and a desire to live with relatives, usually in another community, are other reasons for the Guarijío to make residential moves.

Gentry notes some "short local migrations" between 1933 and 1938, when several households moved from one community to others. Gentry suggests the depletion of wild resources around a community may account for these moves (Gentry 1963:74). The Guarijío may be more dependent than the Rarámuri on wild foods to supplement agricultural produce (W. Dodd, personal communication, 1989), which would supplement Gentry's explanation for the moves.

One other variation in Guarijío settlement has been recorded. This was documented by Jay Rowed, of the Summer Institute of Linguistics, who traveled to the communities of Chagayvo and Cachavachi in 1947. Rowed describes a pattern of pastoral transhumance wherein Guarijío abandon their lowland communities after planting and move to other residences in the mountains while pasturing their goats (Rowed 1947:folio pp. 16-18, as taken from W. Dodd, personal communication, 1988). Neither Gentry nor Dodd observed this mobility pattern among the Guarijío. As with the Rarámuri, however, it is reasonable to assume regional variation in subsistence-settlement strategies, and that Rowed's comments are accurate also.

Guarijío Space Use and Activity Structure

Dodd's ethnoarchaeological work centers on sedentary Guarijío, that is, those who did not make residential moves. The observations, residences, and findings reported here come from these "annually fixed" households. Table 6 lists the activities and their locations given by Dodd in several of his preliminary publications. Figure 42 uses the same format as Figure 36 to identify the important spatial components of the Guarijío residence.

The Guarijío Structural Activity Area

The Structural Activity Area is defined as the core of the residence where the major structures of the site are located. These structures are the kitchen, sleeping/storage structure, and shaded veranda. The structures are usually three distinct enclosed or semi-enclosed spaces, but share one or more walls. The Structural Activity Area has a high use frequency, and consequently is well maintained, with debris from activities cleared and swept up on a regular basis.

Like the Rarámuri, the Guarijío depend on their own produce and animals for much of their subsistence. While food processing activities between the Guarijío and the Rarámuri are quite similar, the location of these activities varies significantly between the two groups. Most Guarijío food processing and preparation activities are confined to the kitchen, and other kinds of activities occur only infrequently here (Table 6).

Facilities typically found in this area include the metate and molino, a firestand for cooking, and a table with chairs and/or stools. The structure also serves as a short term food storage area, and shelves are built for this purpose (W. Dodd, personal communication, 1989).

Although the Guarijío can and do change the placement of their table and chairs, the positions of both grinding facilities (the metate and molino) and the cooking facility (the firestand) are relatively fixed. Dodd describes the set up of the metate in the following way:

> Most Guarijío residential groups have one regularly used metate set on a trifurcate post in the kitchen. The forked top of the post consists of three branches which intersect to form a crotch, into which the metate is seated at a slanting angle roughly 0.6-0.7 meters above the ground. The thick trunk base of the post is set vertically and deeply into a posthole, and this facility remains in the same spot for years. The posts are periodically replaced as they wear out from ground contact; the metates endure for decades or even more than a century. A few households mount metates on massive stone or adobe brick platforms which are also elevated. (W. Dodd, personal communication, 1988)

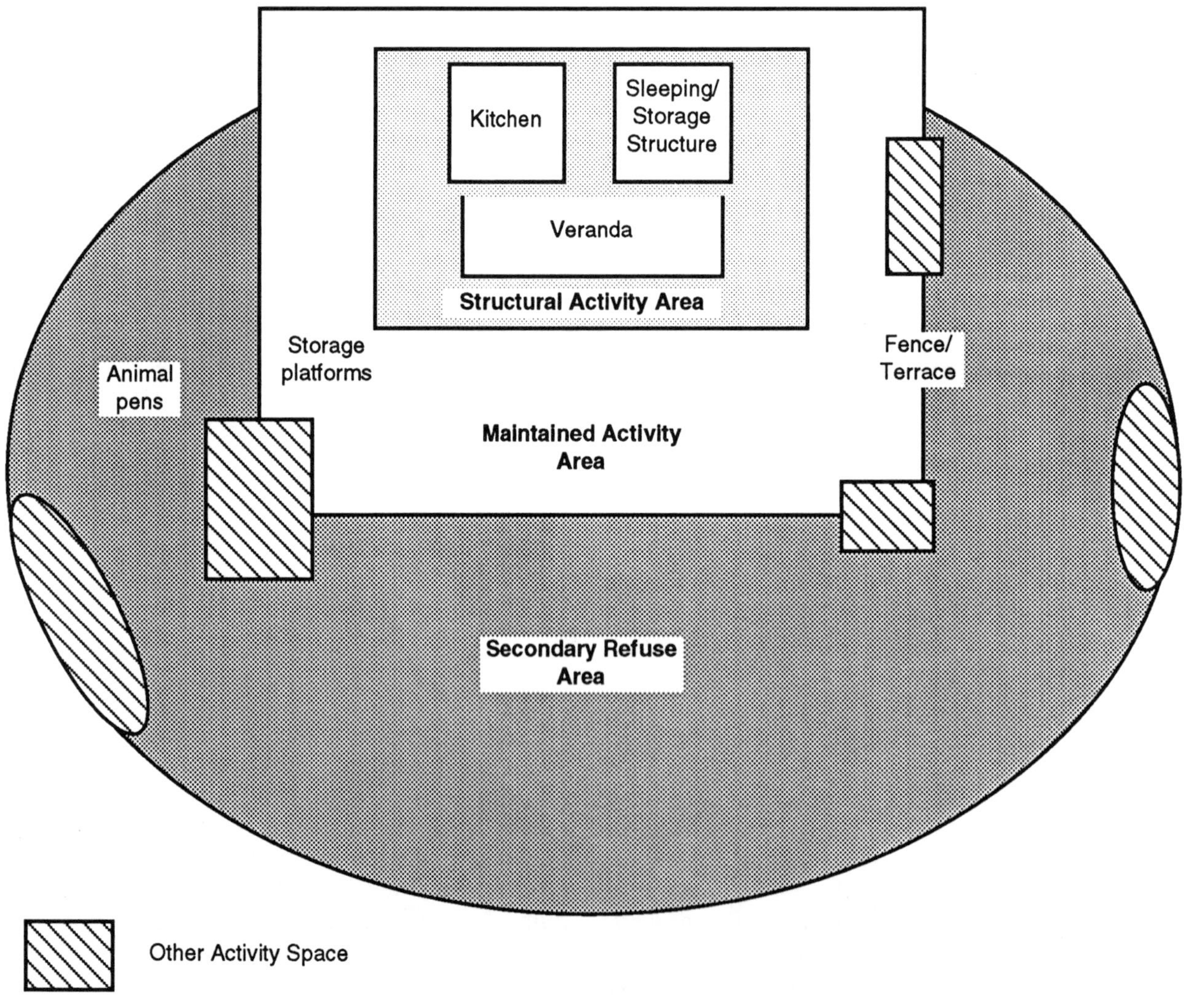

Figure 42. Spatial components of the Guarijío residence.

Molinos are also set up on a post. The Guarijío also set up posts to hold jars of drinking water and other containers. These facilities are found both within structures and outside (Gentry 1963:Figure 12, p. 112). Dodd classifies eight different kinds of fire-related facilities at Guarijío residences, but most common in the kitchen is the firestand. The firestand is also a fixed facility, usually moved from its location only during major renovation.

> The firestand is a raised cooking platform, commonly made of stones mortared with mud.... The firestand is surmounted on top by an open-faced firing enclosure, and usually partitioned into wide and narrow chambers. A sheet metal griddle is laid over the wide side, chiefly for preparing tortillas. Ollas and other utensils are placed to boil on the narrow side or at the open edge of either chamber.... (Dodd 1987:2)
>
> ...Firing facilities rarely occur within the confines of house walls. Firestands, however, do show up within partially or fully enclosed kitchen spaces. They abut or lie very close to adjacent walls, and there is a common tendency to place them in corners. (Dodd 1987:3)

Foodstuffs and containers are left in short-term storage in the kitchen. These items are located as far from the cooking facilities as possible, thus minimizing their risk as fire hazards (Dodd 1987). This segregation also separates cooking from other food processing and preparation activities, and minimizes the overlap of tasks in a limited work space.

The veranda, an unwalled, roofed area, provides a sheltered work place during the day and a protected sleeping place at night. This space is often used for activities that require both shelter and light. Shelling corn and sorting beans are food processing activities carried out on the veranda that do not require facilities (Table 6). Dodd emphasizes the lack of spatial constraint by referring to these as "mobile" activities (Dodd 1984:6). Here, I use the term "movable activities." Movable activities occur in the shelter provided by the veranda and other parts of the Structural Activity Area, and throughout the Maintained Activity Area (see below). Activities carried out under the veranda are also performed in a number of other places throughout the residence, and inhabitants choose the veranda itself when certain temperature and lighting conditions are met (Dodd 1984). These activities are also performed in the kitchen on occasion, but usually not in the sleeping/storage structure (see below). The lack of light within that building plays a significant role in minimizing the activities occurring inside.

In the mild climate of the area, sleeping outside poses no problems for individuals throughout most of the year. The Guarijío sleep either on the ground or on a raised platform. Even during the winter, the Guarijío sleep outside under the veranda (temperatures are seldom below 40 degrees Fahrenheit, or about 4 degrees centigrade, at night), but they will relocate to be nearer a hearth or brazier (containers for warm coals). On especially wet or windy nights they will sleep in the kitchen or the sleeping/storage structure (W. Dodd, personal communication, 1986).

The veranda is kept clear of non-movable facilities. Braziers and unlined fire basins are used when necessary for warmth but braziers are portable, and fire basins are only temporary hearths, without formal preparation or maintenance (Dodd 1987).

The sleeping/storage structure is rarely used as a dwelling. During inclement weather the Guarijío may sleep in it, but the structure's principal use is storage. In a time-motion study, 71-86 percent of the exits and entries into these structures were related to storage tasks (Dodd 1984).

Despite its primary storage function, anthropologists often refer to it simply as the "house," probably to stress that it does not function as the "kitchen" (Dodd 1984, 1987; Gentry 1963). These structures are without windows and so have very little light inside. The Guarijío perform few activities inside these buildings.

These sleeping/storage structures contain a corn crib or bin that is seasonally assembled and dismantled (W. Dodd, personal communication, 1988). According to Gentry (1963:110-111), the storage crib fills from one quarter to one third of the structure, and is slightly elevated above the ground. In addition to storing corn, the structure provides storage space for most classes of belongings at the residence, including clothing, ornaments, tools, containers, musical instruments, and bedding. Gentry reports that, "In [the sleeping/storage structure] are kept the few valued possessions, as clothing, rattles, violins, and very often maize" (1963:112).

In contrast to the Rarámuri, who subsist year-round on stored corn, Guarijío stores of maize last approximately five months. The Guarijío consume more green corn than do the Rarámuri. They incorporate more wild foods into their diet, as well. Both Guarijío and Rarámuri store maize on the cob. The Guarijío shell maize and put it into bags when nearing the end of their supply (in early May), after

Table. 6. Guarijío activities and their locations.

Activity	Location	Facility	Other	Reference
cooking	kitchen	firestand		Dodd 1987:Fig. 3,4
cooking(?)	kitchen, patio	trivet hearth		Dodd 1987:Fig. 3,4,6
cooking	kitchen	log hearth		Dodd 1987:Fig. 6
cooking	patio/edge of Maintained Activity Area	chambered ground hearth		Dodd 1987:Fig. 6
cooking	kitchen	firestand		Dodd 1987:Fig. 1
cooking	patio	?		Dodd 1987:Fig. 1
cooking	kitchen, out of doors	groundstone, molino	locate activity at facility	Dodd1984:3-4
cooking	out of doors	tertiary hearth	locate activity at facility	Dodd1984:4
grinding	kitchen	molino, metate	locate activity at facility	Dodd1984:3-4, Fig. 1
eating meals		table	locate activity at facility	Dodd1984:4, Fig. 1
shelling corn			movable activity	Dodd1984:5-6, Fig. 3
culling beans			movable activity	Dodd1984:5
pit butchering	out of doors	secondary hearth	locate activity at facility	Dodd1984:4
cook *chicharrones*	out of doors	tertiary hearth	locate activity at facility	Dodd1984:4
dispatching goats	beyond central living space		'unpleasant/ messy' activity	Dodd1984:9
primary butchering (goat)	beyond central living space		'unpleasant/ messy' activity	Dodd1984:9
cleaning goat viscera	beyond central living space		'unpleasant/ messy' activity	Dodd1984:9
dispatching chickens	beyond central living space		'unpleasant/ messy' activity	Dodd 1984:9
plucking chicken feathers	beyond central living space		'unpleasant/ messy' activity	Dodd 1984:9

Continued on next page

Table 6—*Continued*

Activity	Location	Facility	Other	Reference
pit-baking agave	beyond central living space	hearth	'dangerous'	Dodd 1984:9
sweeping	kitchen, veranda, patio	broom		Dodd 1984:10, Fig. 6
dumping (ash)	beyond patio			Dodd 1987:Fig. 4,5,6
dumping (ash)	ash dumps in Secondary Refuse Area			Dodd 1984:10, Fig. 6
dumping	beyond Maintained Activity Area, Secondary Refuse Area (kitchen, veranda, patio)			Dodd 1984:10, Fig. 6
ceramic firing	patio, extreme limits of site	hearth		Dodd 1987:Fig. 5,6
firing large numbers of ceramics	beyond central living space	hearth	'dangerous'	Dodd 1984:9
storage	patio	storage structure, bin, crib	other than 'house'	Dodd 1987:Fig. 4,5
storage	sleeping/ storage structure	structure		Dodd 1984:8, Fig. 4
storage	kitchen			Dodd 1984:8, Fig. 4
sleeping	patio, veranda, house, kitchen	mats or elevated beds		Dodd 1987, Fig. 2
twisting rope	patio		movable activity	Dodd 1984:5-6, Fig. 2,3
sewing	patio			Dodd 1984:5-6, Fig. 2,3
woodworking			movable activity	Dodd 1984:5-6, Fig. 3
weaving baskets			movable activity	Dodd 1984:5-6, Fig. 3
washing clothes	boulders and rocks		locate activity at facility	Dodd 1984:3
chopping wood	beyond central living space		'dangerous'	Dodd 1984:9
urination/defecation	beyond central living space		'personally offensive'	Dodd 1984:9
delousing	beyond central living space		'personally offensive'	Dodd 1984:9

which they may dismantle the corn crib inside the sleeping/storage structure (W. Dodd, personal communication, 1989). This more limited dependence on stored maize is probably one reason why, unlike the Rarámuri, the Guarijío do not have separate, permanent storage structures.

The Guarijío Maintained Activity Area

Most Guarijío residences are excavated back into the steep slopes of the mountain and bounded by rock walls and fences. The Maintained Activity Area is the generalized activity space surrounding the Structural Activity Area, but within the confines of the fences and other residence delimiters. The area is relatively open and free of permanent facilities.

As mentioned above, a number of activities performed at the Guarijío residence do not require a facility, and their distribution overlaps with the movable food processing activities. Movable activities observed by Dodd (1984, 1987) include the manufacture of rope, wood working, sewing, and basket making. The locations of these activities vary, depending on a number of factors, including temperature, light, and wind conditions. Spatially extensive activities are carried out away from the core of the residence, but the Guarijío perform the majority of movable tasks within the Structural Activity Area and the Maintained Activity Area.

Along the margins of the Maintained Activity Area, the Guarijío perform tasks classed as unpleasant, unusual, or secondary to daily activities, locating any facilities required by such activities in these places (Dodd 1984). For example, the Guarijío cook outside when large meals are to be prepared or extensive cooking is to be done. In these cases, additional hearths are located along the margins of the Maintained Activity Area or within the Secondary Refuse Area (see below). Killing and processing animals (pigs, goats, and chickens) is often in proximity to these outdoor cooking facilities (see Table 6).

Activities requiring fixed facilities will occur at the loci of those facilities (Dodd 1984:3-5). The Maintained Activity Area has both fixed activity loci (where the activity is dependent on a "permanent" facility), and a generalized activity locus (where the activities are movable or do not require a facility). Some facilities are incorporated into natural features (e.g., boulders used for washing clothes, or the contact point between slope and excavated site boundary used as a hearth location). As in the Structural Activity Area, a number of posts in the Maintained Activity Area hold jars, and other containers. Elevated storage platforms for foods, pots, and other goods are built in the Maintained Activity Area. Gentry (1963:111) describes an elevated platform used for short-term storage at which food and containers are reasonably handy, but away from dogs and other scavenging animals, and also removed from active work space. These facilities may serve as work areas as well as storage space; unlike the Rarámuri, the Guarijío rarely use the ground as a work surface (W. Dodd, personal communications, 1988 and 1989).

The Guarijío Secondary Refuse Area

The Guarijío Secondary Refuse Area is where trash is put once it is moved from the location where it was produced. While the main focus of activities at the Guarijío residence is within an area bounded by fences, rock walls, and the extreme terrain, the Secondary Refuse Area is beyond these limits. The Guarijío locate gardens and build animal pens at contact points between the Maintained Activity Area and the Secondary Refuse Area, or sometimes in one or the other of these areas.

The Guarijío also perform a number of activities at some distance from the Structural Activity Area (Table 6). These tasks include firing large numbers of ceramics, chopping wood, agave pit-baking, animal processing, burning lime for use in preparing corn, delousing, and urination and defecation. Some of these tasks occur as close to the Structural Activity Area as the boundary between the Maintained Activity Area and the Secondary Refuse Area, but many of them are in the Secondary Refuse Area proper, and others are at the far limits of this area. Facilities for extensive and/or intensive firing (for example, the pit baking of agave, the firing of pots, or the burning of lime for processing corn), are usually at a distance from the Structural Activity Area. Other activities not requiring facilities (for example, killing and butchering animals, and urination and defecation), also occur at these extreme limits of the site. Many of these activities are space-extensive, requiring participants to stand and walk around the activity focus, rather than sitting while accomplishing the task (see Binford 1983a:165-172). In intensively used space of any residence, these space-extensive activities are predictably located at the perimeters of the site (e.g., Binford 1978, 1983a, 1987; Chang 1988; O'Connell 1987; Yellen 1977).

Guarijío Site Maintenance

Guarijío regularly maintain their residences; both the Structural Activity Area and the Maintained Activity Area are kept clean of debris in anticipation of upcoming activities. The Guarijío deposit debris from these clean-up activities in the Secondary Refuse Area. Dodd has outlined several contexts for maintenance, including sweeping, dumping, and picking up (1984).

The most comprehensive maintenance activity is sweeping. The Guarijío use small brooms to sweep the ground surface, working from the center of the Structural Activity Area outward. They then deposit the accumulated trash in the Secondary Refuse Area. Because of the systematic sweeping, primary refuse from activity areas is commonly associated with specific areas in the Secondary Refuse Area (Dodd 1984:10). Much of the space on the Guarijío residence is given over to a generalized work place. The only specialized activity areas are within structures (the kitchen and the sleeping/storage structure) and, perhaps, the fixed facilities in the Structural and Maintained Activity Areas. The distribution of debris in the Secondary Refuse Area may reflect maintenance strategy in that different areas within the Secondary Refuse Area associate with particular maintenance zones in the Structural Activity Area and the Maintained Activity Area. At this time, the data are insufficient to assess the implications of whether or not these areas are significantly different from each other. At this time we do not know if they can be associated materially with activity areas within the center of the residence.

Dumping is also a major form of maintenance at the Guarijío residence. Material collected during sweeping episodes is dumped in the Guarijío Secondary Refuse Area. The Guarijío have specific ash dumps in this area where they dispose of debris when cleaning out both indoor and outdoor hearths; Dodd recorded ash was dumped in the same locations 70-90 percent of the time (1987:7). The placement of ash also depends on whether the debris is still smoldering (an effort is made to not ignite other refuse), or if the ash is to be used as an insecticide (Dodd 1987).

The Tuxtleco Campesino

Thomas Killion (1987) carried out an ethnoarchaeological study of the Tuxtleco *campesinos* (peasants) within the Tuxtla mountains of southern Veracruz (Figure 43). His study considered the role of the house lot (residence), in the context of intensive and extensive agricultural practices. Killion's study is directly relevant to the question of how mobility influences the organization and use of residential space. Philip J. Arnold III (1988) also carried out an ethnoarchaeological study among pottery producers of the Tuxtlas. His findings supplement Killion's in the discussion here.

For more than 2500 years, the Tuxtlas region of southern Veracruz has been an area of intensive agricultural production (Killion 1987; Santley 1985). Characterized by diverse topographic and vegetative conditions today, the Tuxtlas region offers both subsistence and commercial agriculturalists and ranchers a rich natural resource base. The Tuxtlas region is an uplifted volcanic area rising above an alluvial plain to about 1700 m. The climate is tropical. The average temperature in the area is 27 degrees Fahrenheit (25 degrees centigrade). Annual rainfall averages 2000 mm, with a dry season lasting from January through May (Arnold 1987).

Tuxtleco *campesinos* occupy their residences year-round as subsistence and semi-subsistence agriculturalists. In addition to the staple subsistence crop, maize, the tropical setting allows for growing a variety of other produce in agricultural, horticultural, and arboricultural settings. Gardens and fields are in production throughout the year. The Tuxtleco not only grow their own staple food, but sell the surplus at nearby markets. The Tuxtleco keep fowl and pigs and, rarely, goats or sheep.

As sedentary intensive agriculturalists, some Tuxtleco cultivate fields that are close to the residence, and others that are farther away. In the latter case, they often overnight at fieldhouses. Killion defines *outfields* as those fields located at more than 40 minutes' walk from the residence; *infields* are within 40 minutes' walk of the residence, and frequently closer (Killion 1987:109). When outfields are a major part of the agricultural system, various duties otherwise carried out at the residence are performed at the outfield and, if it is quite far from the residence, its associated field house.

> Where fields close to the residence are intensively cultivated...the household residential lot functions on a daily integral part of farm production activities. On the other hand, where households focus their productive efforts on more distantly located fields, the orientation of the productive efforts of the household no longer center exclusively on the primary residence but

at more remote locations. (Killion 1987:51-52)

In other words, the Tuxtleco have a direct and measurable relationship between residential site structure and subsistence-settlement organization. With increased intensification of cultivation at infields, residential sites show a greater specialization of activity areas.

Tuxtleco Space Use and Activity Structure

The spatial components of the Tuxtleco residence are the Structural Activity Area, the Maintained Activity Area, the Concentrated Refuse Area, the Garden with Diffuse Refuse, and the House Lot Border (Figure 44). Both the figure and the discussion that follow are adapted from Killion (1987, especially pp. 208-215 and 223ff). Although the spatial components described here are the same as Killion has defined (see below), I have renamed the areas for the sake of consistency within my study.

In discussing the organization and use of residential space among both the Rarámuri and the Guarijío I have described the areas and facilities found at residences, and the way activities are "mapped on" to these. Killion's study focuses primarily on agricultural intensification and overall site structure, rather than discrete activity loci at the residence. While the following discussion of activities at Tuxtleco residences lacks detail, it nevertheless provides some interesting comparisons with the data already presented.

The Tuxtleco Structural Activity Area

The Structural Activity Area is comprised of the dwelling itself and its surrounding area, especially that sheltered by the eaves of the structure's

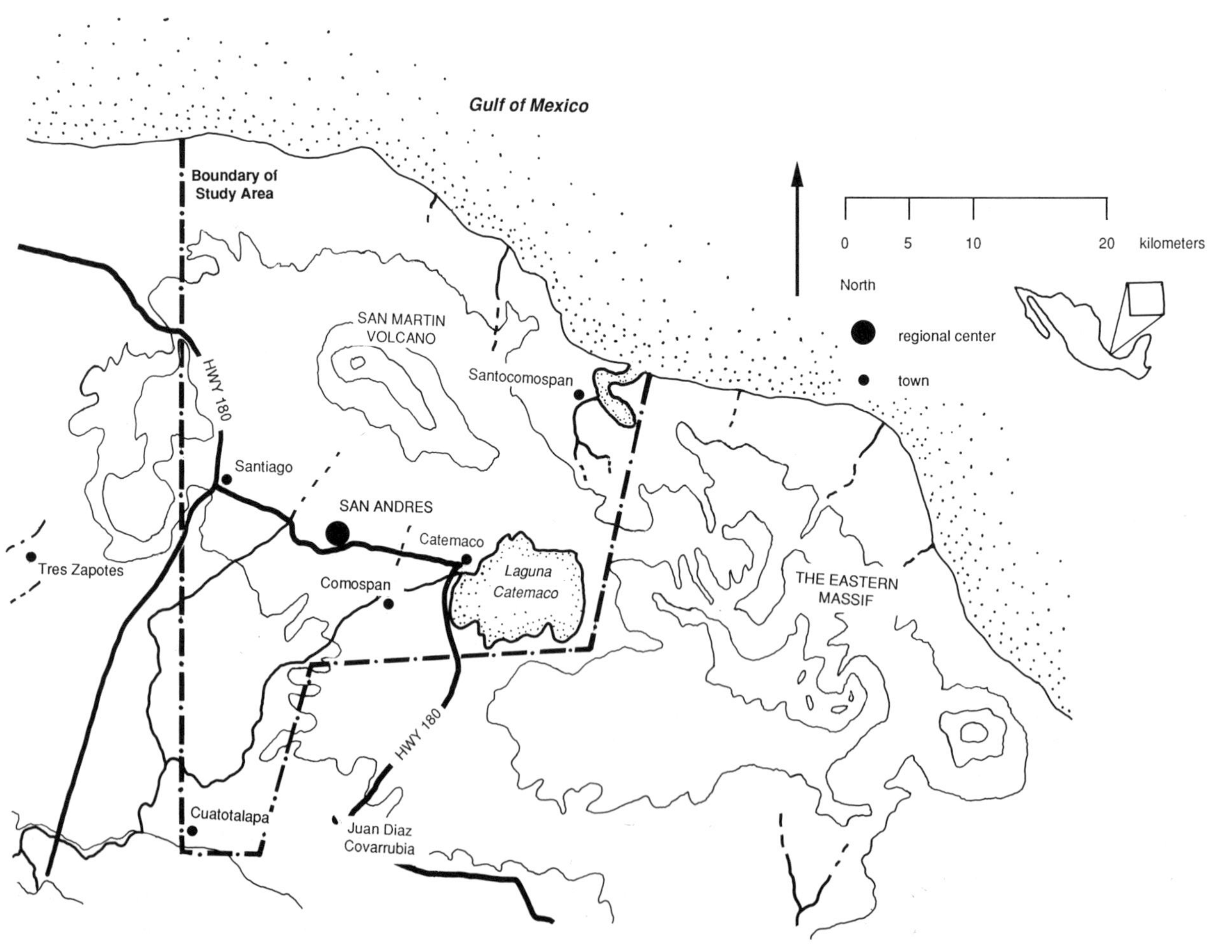

Figure 43. Location of the Los Tuxtlas study area (after Killion 1987:Figure 1).

roof. The Structural Activity Area on the Tuxtleco house lot refers to Killion's "Structural Core" (Killion 1987:224). This area is the most intensively used space at the residence, being the locus of food processing and preparation, leisure, craft and tool manufacture and maintenance, sleeping, and many storage activities.

The Tuxtleco divide the interior of the dwelling into two parts, for cooking and sleeping (designated 1 and 2 in Figure 44). Killion describes this division of space as follows:

> Sleeping and cooking areas are found within two separate zones that are spatially opposed and in many cases placed in two different structures. The activities conducted within each of these two zones are mutually exclusive; that is, people do not normally cook or prepare food in the same living space within which they regularly sleep. (Killion 1987:224-225)

The cooking and sleeping areas in the Structural Activity Area are either separated within a dwelling, or are two separate structures.

The dwelling functions as a major storage facility at the residence, with "tools, agricultural produce, clothing, and domestic utensils" (Killion 1987:225) sorted inside the sleeping area. Killion's description of the interior of the residential structure suggests it is full of items, activity areas, and people performing activities. Clear spaces in the dwelling serve as activity loci and access routes from one area to another within the dwelling.

The Tuxtleco perform the majority of their food preparation activities, including cooking, within that part of the Structural Activity Area referred to as the "cooking area." Killion refers to a raised hearth (a built facility) inside the dwelling on which presumably most of the cooking is done (1987:211). Grinding activities may also be located inside the structure. Outdoor hearths are used occasionally for large food processing projects. The Tuxtleco shell maize, hull beans, and perform similar tasks within the Maintained Activity Area (see below).

The Structural Activity Area extends beyond the interior of the dwelling to the sheltered space outdoors, under the dripline of the roof of the structure. This area contains the same activities and materials found inside the house. There is little differentiation in actual use of indoor and outdoor space in the Structural Activity Area except with regard to some food processing and cooking activities (Killion 1987:422).

Because the tropical climate remains mild much of the year, only during the rainy season is sheltered activity space critical. The family sleeps out of doors much of the year, using the sleeping area primarily during the rainy season. Throughout the entire year, the sleeping area provides storage space for food, clothing, tools, and utensils. Storage facilities per se within the Structural Activity Area are virtually unknown, but shelves, roof beams and posts, wall nooks and crannies, and benches within and outside of the dwelling are drafted into use as storage space. This space is also used on occasion as an area for receiving guests, and as private space for household members when they wish to remove themselves from the rest of the family.

The Tuxtleco Maintained Activity Area

The Maintained Activity Area is the frequently used activity area surrounding the Structural Activity Area. Killion refers to it as the "Clear Area" (Killion 1987:227). Activities occurring within the Maintained Activity Area are perhaps of shorter duration and occur less often than many of the activities performed within the Structural Activity Area. The Tuxtleco place additional storage structures for agricultural and other subsistence activities in the Maintained Activity Area, but the majority of these storage facilities are presumably located in the Diversified Activity Zone (see below).

The Maintained Activity Area is divided into two parts—the Staging Zone and the Diversified Activity Zone (3 and 4, Figure 44). These zones are spatially as well as functionally distinct. The Staging Zone is that part of the Maintained Activity Area associated with activities that, in the course of their duration, pass beyond the limits of the residence. These activities are actually accomplished away from the residence, but are prepared for, or terminated there.

> These activities include the preparation of cultivating tools for transport and use, arranging packs and loads for transport, temporary storage of items to be transported to distant locations, and other activities that produce little production debris. (Killion 1987:229)

The Tuxtleco also locate religious activities and celebrations in this zone. Play and leisure activities easily extend from the Structural Activity Area into this space.

The Staging Zone is well-maintained and intensively used on a daily basis. The nature of the activities occurring in this zone, however, is such that they generate little debris. Access to the area beyond the House Lot Border (see below) is immediate from this part of the residence. There is little in the way of refuse along the perimeter of the zone. Within this zone, but not in competition with activity space, are "provisional dumps" and storage areas where potentially useful items are accessible.

In contrast to the Staging Zone, the Diversified Activity Zone is "...an area utilized for small-scale industrial activity, food processing, socializing, and leisure" (Killion 1987:230). Activities in this zone include "carpentry, tool repair, fish net mending and construction, boat building, and ceramic manufacture...[and] drying and shelling maize and beans..." (Killion 1987:231). These activities generate diverse amounts and kinds of debris.

Activities within the Diversified Zone typically are of variable duration. Some occur many times a day; others, lasting a number of days or weeks, are placed in one part of the activity zone where they will not be disturbed by other tasks. This zone is more casually maintained than is the Staging Zone, but there is a greater amount of refuse both in and surrounding this space.

Killion's study of Tuxtleco *campesinos* considers the role of the residence in the context of the agricultural strategies of communities with different environmental resources. In a system using both infields and outfields, there is an important relationship between the organization and use of Tuxtleco residential space and the degree of reliance on and intensity of use of the two types of fields:

> As the infield focus of agricultural production increases, so does the size of the [Maintained Activity Area] within the lot.... Conversely, the size of the [Maintained Activity Area] decreases as the intensity of cultivation on outfield plots increases, and...this relationship is probably curvilinear.... Thus, it seems likely that as the intensity of cultivation increases on outfield plots, more activities are conducted at the remote locations than would normally have been conducted around the primary residence. The shift in activities reduces the amount of [Maintained Activity Area] necessary for agriculturally oriented activities within the house lot, and in response these spaces tend to shrink in size. This reduction in space will especially affect the size of the staging zone. (Killion 1987:341)

Thus, the size of the Maintained Activity Area in general, but the Staging Zone in particular, is sensitive to the amount of land held in infields and outfields. With increased occupation of the residence (for example, for those residences over thirty years old), this relationship is less well-defined. Killion concludes that agricultural intensity as measured by intensive and extensive reliance on infields and outfields results in meaningful variability in residential site structure.

Arnold's work among the potters of Los Tuxtlas supports this compartmentalization, expressed in terms of areas that are both activity specific and activity exclusive (Arnold 1987, 1988). Although not in the same study communities as Killion's, and having a different economic orientation, Arnold's site structural study of ceramic production on Tuxtleco house lots shows that intensification of activities into smaller areas leads to specialization of that space (Arnold 1987, 1988, 1991). With decreased size of the house lot, Tuxtleco ceramic producers switch from open-air firing to the use of kilns. This shift is a response to scheduling pressures for work space. As the area available for activities decreases, activity areas become more activity specific and more activity exclusive (Arnold 1987:7). In the process of transformation from more general to more specialized work space, behavioral organization changes first (Arnold 1988). A frequently performed activity is likely to occur in a specific place, eventually to the exclusion of other activities in that place. When pressure for work space continues, a change in technological organization results. Arnold refers here specifically to the change from open-air to kiln firing among different communities of potters in the Tuxtlas region. While these communities have a different economic base than do the Tuxtleco of Killion's study, the residential site structure for the region in general remains quite similar. More importantly, the Tuxtleco from the two study groups consistently use the same strategies in dealing with limitations in residential space.

The Tuxtleco Concentrated Refuse Area

The Structural Activity Area is the most intensively maintained area at the residence. In addition to daily sweeping in this space, other forms of maintenance include the timely replacement of items to their place of storage after use, and piling refuse in out-of-the-way places to be dumped later in the Concentrated Refuse Area. The Maintained Activity Area is kept clean in readiness for activities and because it is a means of access both into and out of the Structural Activity Area. As part of residential maintenance, the Tuxtleco occasionally set fire to refuse concentrations. They also stack piles of discarded objects at the boundaries of the area. Some of the piles of refuse are sorted and serve as raw material resource areas. With increased distance from the Structural Activity Area, Killion reports a decline in both activities and maintenance within the houselot. The Concentrated Refuse Area has no clear-cut boundary; rather, the density of trash increases as one enters this area from the Maintained Activity Area.

The Concentrated Refuse Area is that portion of the residence in which refuse from the Structural Activity Area and the Maintained Activity Area is deposited. This area is identical to Killion's "Intermediate Area" (Killion 1987:232). The Concentrated Refuse Area encircles much of the site, but is negligible to non-existent in front of the Staging Zone. The Tuxtleco maintain paths transecting the Concentrated Refuse Area from the Structural Activity Area to the House Lot Border in much the

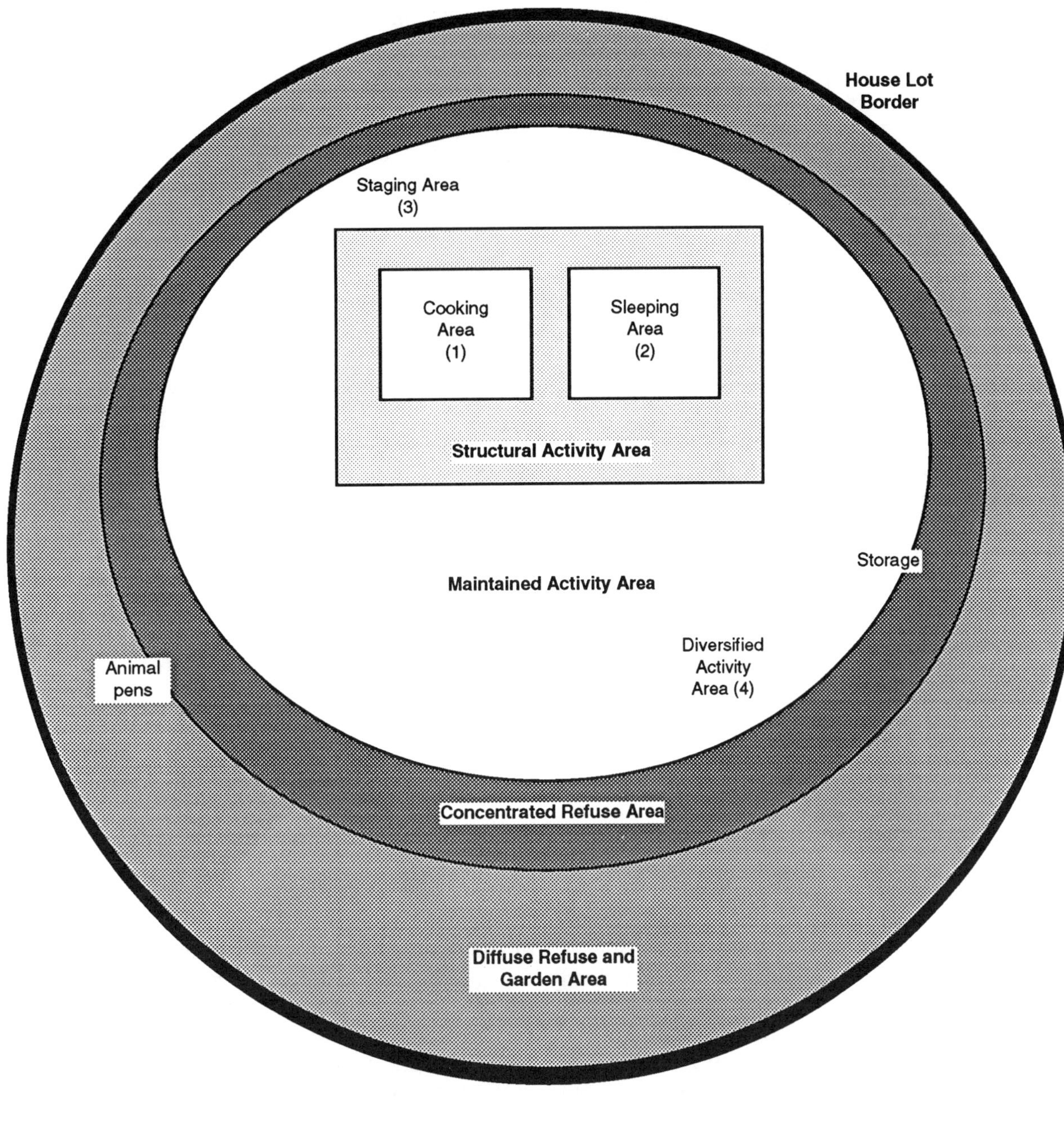

Figure 44. Spatial components of the Tuxtleco residence.

same way as they do the Maintained Activity Area. The routes themselves are clear, with small piles of refuse located off to one side. Animal pens are usually in the Concentrated Refuse Area or at the contact points between the Maintained Activity Area and the Concentrated Refuse Area. The Tuxtleco treat these spaces as inappropriate for any human activities other than urination and defecation. Most or all animal pens are in the area around the Diversified Activity Zone rather than the Staging Zone.

Sweeping is one of the major forms of maintenance by which trash accumulates in the Concentrated Refuse Area. Household members will also collect materials from the activity areas and dump them here. The Concentrated Refuse Area contains a wide variety of artifact types, but the processes of decay in the tropical climate and the activities of animals serve to strip away most of the organic matter, creating a ring of hard, durable inorganics around the Maintained Activity Area. The Concentrated Refuse Area in older house lots becomes a raised midden.

Killion includes a measure of concentrated refuse piles (referred to in his analyses and text as "REF") as an important indicator of refuse treatment. These are defined as "concentrated piles of hard refuse...found scattered around the lot especially along the boundary of the clear and intermediate areas" (Killion 1987:296). Killion's references to "concentrated refuse piles" should not be confused with my discussion of the Concentrated Refuse Area. Concentrated refuse piles are products of a type of maintenance behavior wherein trash is collected into piles rather than swept to the interface between the Maintained Activity and Concentrated Refuse Areas. The former type of maintenance results in a less amorphous Concentrated Refuse Area than is found with the latter behavior.

For the Tuxtleco, the relationship between agricultural intensity and residential site structure is expressed not only in the size of the Maintained Activity Area, but also in the maintenance behavior at the residences. Specifically, with increased agricultural intensification, the Tuxtleco increase their use of concentrated refuse piles at the residence. As the measure of concentrated refuse piles goes up, the areal extent of trash areas decreases (Killion 1987:346-347). By using this maintenance strategy, the Tuxtleco free up space for use as activity areas when space is limited (defined by the House Lot Border and constrained by adjacent residences) and the need for workspace is great.

The Tuxtleco Diffuse Refuse and Garden Area

The Diffuse Refuse and Garden Area, referred to by Killion simply as the "Garden Area," is an area "...of plants that are cultivated in close proximity to the residential structures and that provide food, fiber, medicine, condiments, construction materials, and fuel (Killion 1987:236)." Refuse spreads within the garden from the Concentrated Refuse Area, and is trapped in and around garden plants (Killion 1987:237). The garden itself plays an important role in the economy and subsistence of the Tuxtleco household; not only are products from the garden consumed by residents of that house lot, but produce is also sold, traded, and given away. The Tuxtleco tend plants in the garden, but beyond paths through this area, they pay little attention to debris accumulating in there. Horticultural activities are the only activities other than urination and defecation that occur in the garden. Debris is neither generated nor disposed of in this area; rather, trash spreads into the garden from the Concentrated Refuse Area as it is dispersed through the traffic of humans and animals at the residence. The Diffuse Refuse and Garden Area surrounds the entire Tuxtleco residence except at the Staging Zone.

The Tuxtleco House Lot Border

The House Lot Border is an alignment of stone, wood, and/or vegetation that encompasses the entire residence. This barrier serves as a boundary marking interior space as personal property. It maintains the privacy of the residence. The House Lot Border defines the limits of the residence, and consequently inhibits the spread of material generated in the house lot beyond this barrier.

Summary and Discussion

As subsistence maize agriculturalists, the Rarámuri, Guarijío, and Tuxtleco share a number of residential activities. Differences in climate, topography, and settlement organization are critical factors resulting in variability between the residential site structure of the groups. This section briefly summarizes the three subsistence-settlement systems and strategies of residential space, and discusses the similarities and differences among them.

Subsistence-Settlement Organization

The Rarámuri and Guarijío are both considered subsistence agriculturalists, growing corn, beans and squash. Both supplement their diet with wild plant foods, and some meat and fish. Wage labor is one way the two groups gain cash to buy various items; currently the data are insufficient to determine if either group is more or less dependent on the Mexican cash economy. The Tuxtleco are also dependent on maize agriculture for the majority of their subsistence, and they supplement this staple with beans, squash, fruits, wild plants, and garden produce. The garden plays a particularly important role in Tuxtleco subsistence as well as in their residential site structure. Tuxtleco gardens and fields are in production virtually year-round, and there is little emphasis on food storage (a major concern at the Rarámuri residence and an important factor in Rarámuri residential site structure). As mentioned above, the Tuxtleco also sell surplus produce at nearby markets.

Both the Rarámuri and the Guarijío keep domestic animals including cattle, goats, pigs, chickens, and dogs. The Rarámuri also maintain small flocks of sheep, or mixed herds of goats and sheep. Rarámuri are required to take their herds out daily and accompany them while the animals graze. Even if the herder is only a four- or five-year-old child, the herd is supervised closely. This is to protect the animals and, during the growing season, to prevent the animals from getting into the fields and destroying the crops. Cattle, too, are herded conscientiously during the growing season.

The incorporation of sheep into Rarámuri animal husbandry may have played an important role in differences in the movement of both people and animals in the Rarámuri and Guarijío systems. The Guarijío from La Barranca and Chiltepín spend less energy in herding activities than do the Rarámuri. Cattle and goats are left to roam free in areas away from fields, or are penned during the growing season (W. Dodd, personal communication, 1988). The thorny vegetation is said to discourage the raising of sheep (W. Dodd, personal communication, 1989). Winters in the Guarijío region are milder than those in Rejogochi. The Rarámuri's biseasonal moves for protecting animals are apparently not necessary among the Guarijío.

The Rarámuri pen animals on portions of fallow fields during warm weather, and in the winter, move the animals to corrals in rockshelters or barns. In the spring, they then carry the accumulated manure from the winter corrals down to fertilize the fields. This practice is crucial to the Rarámuri agricultural system; without fertilizer, not only are fields less productive, but they must be put into fallow more frequently. The Guarijío practice of letting animals roam without supervision raises the possibility of other differences from Rarámuri agricultural practices and mobility strategies. Exploring these differences must wait until the land use practices of the Guarijío regarding field sizes, productivity, and fallow cycles are better known.

While pigs and chickens are present at the Tuxtleco residences, most households do not have herds of sheep or goats. Currently, no additional details are available on the role of domestic animals in Tuxtleco economy and residences.

The Rarámuri differ in several important ways in their settlement organization and mobility strategies from those of the Guarijío and Tuxtleco. The Rarámuri are residentially mobile during the agricultural cycle, as well as making biseasonal moves from cold weather to warm weather dwellings. The Rarámuri may change the emphasis (function) of residences after a few years, as well. The majority of the Guarijío are residentially sedentary both throughout the year and over the course of a number of years, although small numbers of Guarijío are residentially mobile during some years. Generally, these instances of residential mobility are for the same kinds of reasons given for the Rarámuri. The Tuxtleco are residentially nonmobile. Within their system of infields and outfields, the Tuxtleco normally tend even distant fields by making day trips and returning to the house lot in the evening. If the outfield is especially far from the residence, the Tuxtleco farmer will overnight at a fieldhouse.

The Rarámuri also tend fields at distances of 40 minutes or more from their residence (the minimum distance of the Tuxtleco outfield from the house lot). But while the Tuxtleco farmer makes day trips or stays overnight at the outfield, the Rarámuri household makes residential moves to fields at the same distance from its main residence. The Rarámuri household will spend a number of days or weeks in the new community not only tending its crops, but also socializing, and assisting its neighbors in this community with various tasks.

Residential Site Structure

Residences for each of the three groups are composed of three main areas—the Structural

Activity Area, the Maintained Activity Area, and one or more types of refuse areas. Structural elements in these areas influence the flow and direction of activities and debris. In the Structural Activity Area, the structures themselves influence the placement of activities both indoors and out. In the Maintained Activity Area, structural elements include topographic features, as well as some buildings and facilities, for example, hearths, posts, storage platforms, and animal pens. Many of the elements in refuse areas, are topographic features and fence or rock alignments.

Rarámuri residences have the least formalized site structure of the three cultural groups. The Rarámuri locate their food processing and preparation, sleeping, and socializing activities within the same space. Among the Tuxtleco and Guarijío, food-related activities are carried out in a separate place, and generally in a separate structure, from sleeping and socializing areas. The lack of formal activity space at Rarámuri homes is most obvious in the differences in facilities' relative permanence and the consequent arrangement of activities and activity areas, and in the specialization of activity areas among the Guarijío and Tuxtleco. With the exception of the storage structure, there is virtually no specialization of Rarámuri residential space. Facilities that are fixed in space among the Guarijío and Tuxtleco remain movable at the Rarámuri residence. These include grinding implements (especially the metate and molino), and cooking hearths. As discussed above, the Rarámuri locate these facilities in any one of several places at their residences. They can always create a new activity area as the need arises. Finally, use of these places is not exclusive to any single type of activity.

The Guarijío Structural Activity Area has at least two distinct specialized activity areas: the enclosed food processing and preparation area (the kitchen), and the specialized storage aspect of the sleeping/storage structure. In the Guarijío kitchen, specialization is due in part to the localization and fixed nature of several critical food processing and preparation facilities required for grinding and cooking foods. The fixed nature of these facilities assures the location of their associated activities in this place; however, the Guarijío will move these facilities and/or activities on occasion (W. Dodd, personal communication, 1988). The exclusion of other activities from a place is important because it alleviates the need to schedule disparate activities in the same workspace. This solution to possible scheduling conflicts is in direct contrast to the Rarámuri strategy, which relies on the relative ease of occupying any of a number of places using movable food processing facilities. With increasingly fixed activity areas in bounded space, workspace becomes more constrained. Activities requiring a facility must be performed at that place. If an activity positioned there does *not* require that facility, participants in the activity must work around it. Movable projects are likely to be ousted whenever a worker needs the facility and that specific activity area. The scheduling of tasks and specialization of work areas minimize many problems of the spatial overlap of disparate activities.

The Guarijío sleeping/storage structure is also something of a specialized activity area. The specific function of the structure as storage space for both food and other goods imposes a number of requirements on the building. Most important, the building must be relatively secure against the elements and various pests—insect, animal, and human. Guarijío structures are of rock, adobe, wattle and daub, or an amalgamation of these construction materials (Dodd 1984; W. Dodd, personal communication, 1988; Gentry 1963). The sleeping/storage structure is completely enclosed, without windows or access to light. The Guarijío do not put hearths or stoves in these structures, and the buildings are without smoke holes or chimneys. The complete lack of light in these structures effectively inhibits its use for many activities.

On Tuxtleco residences, activity structure and placement differ in three basic ways from those of the Rarámuri. First, the Tuxtleco perform the majority of their activities in the two zones of the "Structural Core" of the house lot; the Rarámuri, while carrying out many of the same kinds of activities, lack the definition of the Tuxtleco Structural Activity Area. Secondly, the Maintained Activity Area in the Tuxtleco residence is more compartmentalized for activities than it is on the Rarámuri residence. Finally, maintenance behavior at the Tuxtleco residence is far more structured than at the Rarámuri residence.

The organization and use of space at Tuxtleco residences is influenced greatly by the need to perform a number of activities in physically constrained space. As scheduling pressure for work space increases, there is a concomitant compartmentalization of activities into specialized work areas (Arnold 1988). This specialization of space extends to the organization and use of refuse areas and is reflected in Tuxtleco maintenance behaviors. The division between cooking and sleeping areas seen on the Guarijío residence is also present at the Tuxtleco home. Tuxtleco outdoor work areas are divided into specific kinds of activity space. These subdivisions for specific types of activities are greater than is the case on either Rarámuri or Guarijío residences. The Tuxtleco Maintained Activity Area encompasses the Diversified Activity Zone and the Staging Zone. Within these zones, especially within the Diversified Activity Zone, there are discrete work places. In this way, the Tuxtleco locate long-term projects in places separate from areas housing short-term activities.

The need for sheltered activity space in response to climatic conditions varies between the three groups. In the Guarijío region, the climate is warmer throughout the entire year than in Rarámuri country. The Guarijío require less sheltered lifespace for warmth during the cold months, and more sheltered lifespace for shade during the warm months. The veranda used by the Guarijío and the unroofed sheltered activity area of the Rarámuri are probably functionally equivalent but reflect climatic variability and the need for qualitatively different sheltered lifespace.

The Tuxtleco live in a tropical environment, with relatively higher rainfall and warmer year-round temperatures, compared to the other two groups. The Tuxtleco have no biseasonal shift from warm weather to cold weather residences; rather, their need for sheltered lifespace comes during the rainy season. As with both the Rarámuri and the Guarijío, outdoor areas make up a major part of the activity space at the residence. Seasonally the Tuxtleco use interior space for most of their activities during the rainy season, however. These conditions are important in influencing residential site structure, specifically in terms of different requirements for sheltered lifespace. In the case of the Rarámuri, shelter requirements are primarily in response to cold and precipitation. The Tuxtleco require seasonal shelter only during the rainy season, when numerous activities are transferred from the Maintained Activity Area into the dwelling itself specifically, into the Sleeping Zone.

Subsistence, mobility, and requirements of sheltered lifespace are all important factors influencing how the Rarámuri, Guarijío, and Tuxtleco organize and use their lifespace. Some of the most significant differences in residential site structure of the three groups are specialized work places among the Guarijío and Tuxtleco for activities the Rarámuri perform in unspecialized areas. The contrast in space use for food preparation activities is one of the most important examples.

All three groups shift many common daily activities between the Structural Activity Area and the Maintained Activity Area at their residences. With distance from the residence core, variation between both the kinds of activities and their placement on Rarámuri, Guarijío, and Tuxtleco residences becomes more evident. The activities occurring in the area between the Maintained Activity Area and the Secondary Refuse Area on Guarijío habitations either are not common tasks, or are performed at these locales as a result of increased competition for workspace within the Structural Activity Area. For example, the processing of an animal for consumption requires additional cooking facilities for extended periods of time. To avoid scheduling conflicts with regular cooking activities, large hearths are placed at the periphery of the Guarijío Maintained Activity Area and in the Secondary Refuse Area. The Tuxtleco divide the Maintained Activity Area into the Staging and Diversified Activity Zones thereby avoiding scheduling conflicts, or problems due to overlapping spatially extensive activities.

The placement of Guarijío activities beyond specialized activity areas is more structured than is the case when Rarámuri locate places to perform the same types of activities. Many tasks that a Guarijío worker might take into the Maintained Activity Area or Secondary Refuse Area, would remain at the interface between the Structural Activity Area and the Maintained Activity Area on Rarámuri residences. An example of this is the placement of ceramic firing activities. The Guarijío remove ceramic firing to the Secondary Refuse Area, whereas the Rarámuri perform the same task at the outdoor hearth within the Maintained Activity Area. Not only does the potter require prolonged use of a large fire, but also space for movement around the hearth to tend the fire, and an area to put the pots while they cool. The Tuxtleco, in dividing the Maintained Activity Area into two zones, have a more specialized use of space. In the example of firing ceramics, Tuxtleco in spatially constrained residential settings have adopted a technological change in facilities—the kiln (Arnold 1991). This change represents an even more specialized use of space than the specialized activity area and facility used by the Guarijío.

The Guarijío use the Secondary Refuse Area as an additional zone for some kinds of activities. Locating activities at the interface between the Maintained Activity Area and the Secondary Refuse Area is a compromise between relative accessibility and proximity to the activities and resources at the core of the residence, and the debris and mess generated by the activity in question. In contrast, the Rarámuri continue to carry out most of their activities within the undifferentiated space of the Maintained Activity Area. One reason for the different placement of identical activities by Guarijío and Rarámuri workers is the consideration of an activity's spatial constraints and debris resulting from its performance. Another reason is the consideration of the *other* activities and spatial constraints occurring within that same space. The Tuxtleco divide their space into areas that are not only activity specific, but activity exclusive, as a way to deal with scheduling and other issues of constrained space use.

Trash and Maintenance

The distribution of trash by the three groups differs with variations in topography and vegetation. Rarámuri residences are often on relatively flat land adjacent to their fields. In this context, the results of different maintenance behaviors are clearly shown. On Rarámuri residences, much of the swept debris stays in the Swept Trash Area. The Guarijío and Tuxtleco demarcate their living space with fences and walls, or the thick vegetation of their gardens. Guarijío residences are also constrained by extremes in natural topography. Households discard their refuse *beyond* the limits of this activity space. Beyond this boundary trash accumulates in much the same manner as in the Rarámuri Swept Trash Zone, but with less subsequent diffusion or scattering back into the residence because of the physical barrier. Additional trash is dumped and tossed on top of this swept and discarded material, as in the Rarámuri Concentrated Refuse Area. At Rarámuri rockshelter sites, the distribution of trash on the slope below the built-out activity area (Figures 28 and 9), may be more similar to the Guarijío trash pattern than at other Rarámuri homes because rockshelter habitations share similar topographical and physical demarcators for refuse areas with Guarijío residences. There are also important differences in maintenance activities and space use by the three groups in refuse areas.

Rarámuri hearth maintenance appears to be less systematic than is Guarijío. Rarámuri infrequently clean their hearths, and the locations of their ash dumps are often variable. Some households, like that occupying Residence 17, consistently use the same dump(s) (Figure 34). Others do not, as the diffuse distribution of ash at Residence 11 shows (Figure 31). In many cases, the ash and charcoal scatters shown on the figures of Rarámuri residences are the locations of hearths, and are not cleaned regularly. Low maintenance is especially common for the outdoor hearths, for example, those shown at Residences 4 and 8 (Figures 28 and 33, respectively). The Guarijío have concentrated dumps of ash and other debris within a generalized refuse area. The reason for differences between Rarámuri and Guarijío hearth maintenance and refuse disposal is the more regular maintenance of hearth facilities necessary with the formal (and activity-specific) hearth facilities of the Guarijío. Guarijío firestands serve to contain cooking fires. These facilities consume more fuel and thus generate more ash than do either the Rarámuri hearths, or the shallow fire basins the Guarijío use in other parts of the residence. Their daily use necessitates the regular cleaning and dumping of the Guarijío firestands.

In her work with Navajos on the South McKinley Mine lease in the Southwestern United States, Klara Kelley notes an increase in ash dumps on Navajo sites through time as more families incorporate the use of wood burning stoves in their residences (Kelley 1986). These stoves consume greater amounts of wood than do traditional fires, and generate greater volumes of ash. As a result, the Navajos cleaned the stoves more frequently, using buckets (rather than cedar bark scoops) to haul away the ash. With a container, it was easier to carry this greater amount of ash farther away from the actual living space. The resulting dumps were at greater distances from their residences than had been the case previously (Kelley 1986:84). The firestands at Guarijío residences require the same formalized maintenance behavior (dumping) as do the Navajo stoves.

Comparable data on hearth maintenance is not available for the Tuxtleco. However, much of the maintenance behavior at Tuxtleco residences is dumping. Stockpiling is another important type of maintenance behavior employed by the Tuxtleco. These maintenance strategies minimize the *extent* of refuse, and the space given over to refuse areas, at a residence.

Curatorial maintenance activities involve returning items to their storage places when they are not in use. These behaviors are not discussed widely, probably because they impact the archaeological record differently than do discard behaviors (but see Binford 1978, 1983a). Like discard maintenance, curatorial maintenance is performed on both interior and exterior space. Specialized activity areas used by a limited number of individuals, or for a particular activity and associated with specialized tool kit, are subject to less curatorial maintenance than is space with a high frequency and diversity of use.

Important differences in curatorial maintenance activities exist among the Rarámuri, Guarijío, and the Tuxtleco. Rarámuri sweep both inside their dwellings and in the Maintained Activity Area. Interior space is less well maintained than the outside areas, however. Interior space is used for a variety of activities, many of which require additional tools or facilities. These items (for example, sewing kits, serving dishes, and utensils) tend to be "put away" (curatorial maintenance) anywhere in the structure. When household members finish eating, for example, they may place the bowl or cup they used beyond the space of any ongoing activity. The Rarámuri use the ground as their primary (and often only) working surface. They use low bed platforms as storage surfaces, as well as sitting and sleeping on them. Tables and other raised activity surfaces are rare in Rejogochi. House interiors in Rejogochi usually are cluttered with objects. The structure is swept, but many possessions are left on the floor, in a form of "ready-storage." Exterior activity areas are kept more clear of materials than are dwelling interiors.

The difference in maintenance between outdoor and indoor space among the Rarámuri resembles that seen between the Staging Zone and the Diversified Activity Zone on Tuxtleco residences. While the Tuxtleco Staging Zone and Rarámuri exterior activity areas are kept clean and empty through regular maintenance (in readiness for any activity), the Diversified Activity Zone and interior of Rarámuri dwellings are cleaned less regularly, and more items remain on the surface of the work area after sweeping. The Diversified Activity Zone is the locus of long-term and on-going activities, and maintenance of this area can be disruptive of those activities.

The Rarámuri do not maintain their activity space with the same degree of intensity as do the other two groups, nor do they participate in various discard/clean-up activities to the same degree. Most Rarámuri maintenance consists of sweeping every week or two. In contrast, the Guarijío generally sweep at least once a day, both inside and outside their dwellings (Dodd 1984), and the Tuxtleco sweep twice daily (Arnold 1991).

Summary

This discussion covers several important relationships between residential site structure and subsistence-settlement strategies. Residences of the Rarámuri, Guarijío, and Tuxtleco share basic structural elements, but we see a number of important differences in the formalized use of space between the three groups. The physical separation of many food processing and preparation activities into a discrete area, the transition from open to kiln-firing of ceramics, and the use of formal dumps and stockpiles are all examples of the consequences of increased intensity of occupation at the residence. The question posed here is, can we use this information to learn more about distinguishing mobile from non-mobile agriculturalists' settlement patterns in the archaeological record? This question is explored further in the following chapter.

Chapter 5

The Archaeological Implications of Agriculturalists' Space Use

Introduction

Any activity requires a certain amount and type of space for its performance. If we can begin to assess the nature of the space required, and how potential conflicts or scheduling pressures in the placement and timing of activities are resolved, then we can identify linkages between the patterning of space use and the subsistence-settlement system. While the same rules govern space use by hunter-gatherers, agriculturalists, and groups with mixed economies, there has been little attempt to consider site structural similarities and differences among such disparate groups. Such consideration would help us to understand more fully the factors important to the organization and use of residential space. In this chapter, I include a few examples of residential site structure from other subsistence-settlement systems to illustrate circumstances under which groups with diverse settlement or subsistence patterns face common problems in how they organize and use residential space, and how these groups apply the same principles of site structure to solve these problems. Understanding why similarities in site structure exist among groups with substantially different economies or settlement practices will help us in an accurate interpretation of the variations we find in the archaeological record.

Until recently, archaeologists using materials found in archaeological contexts to assess whether mobility plays a role in the settlement organization of agriculturalists found little of relevance in the site structural literature to inform them on this issue (but see Hitchcock 1982, 1987; Kent 1987). As the rules and relationships of spatial organization are defined and explicated, site structural analysis will become a critical way to identify meaningful variability in space use and to interpret that variability in a broader context.

Archaeological Implications for Site Structure

In this study, my primary focus has been on developing a better understanding of the arrangement of materials in space in order that we can address questions such as: What influences the organization and use of space at a place? How are activities and facilities distributed within that space? How is space use manifested in material terms that we can identify in the archaeological record? It is clear archaeologists can use an understanding of residential site structure to identify meaningful variability in subsistence-settlement systems, and to interpret that variability more accurately. Nevertheless, we face numerous complications when making inferences about the past based on the distribution of selective material remains. Given the findings of the current study, this section suggests some productive approaches to the archaeological record.

Before I discuss these approaches, however, I think it is useful to underscore an important aspect of archaeological analyses I have *not* considered. I have specifically avoided in-depth investigations into the nature of the material record itself in this study. I am not concerned here with how variability in ceramic assemblages, floral and faunal remains, or the presence of other artifact classes might inform us about the mobility strategies and settlement organization of the Rarámuri. Clearly, these are important lines of inquiry. Analyses of these materials should be used together with site structural analyses when making inferences about prehistoric cultural systems. Nevertheless, the exclusive investigation of residential site structure presented here demonstrates that variability in the organization and use of space itself is relevant to many of our basic questions about cultural systems. Admittedly, the "archaeological record" presented here is simplistic. I have excluded any discussion of subsequent occupation beyond the annual or near-annual reoccupations already described. I have not dealt with the wholesale "borrowing" of structural and material elements from sites. Nor have I considered the effects of taphonomic processes on materials and their arrangement. These processes and events certainly are critical in the development and evolution of the archaeological record at any site, and must be considered when dealing with actual prehistoric materials.

While recognizing my neglect of these aspects of archaeological inquiry, I suggest that the importance of the present study is to contribute to the baseline understanding of residential site structure. With such a baseline, we are in a better position to assess how variations in spatial arrangements of features and artifacts reflect settlement organization, and when they might be due to other circumstances.

The Material Record at Residential Sites

The distribution of artifact classes in time and space plays a critical role in archaeologists' interpretations of both a site and the arrangement of sites across a landscape. Ethnoarchaeological studies of site structure among hunter-gatherers and agriculturalists have shown that the meaning of the spatial patterns encountered by archaeologists is not always clear.

Many ethnoarchaeological analyses of space use at hunter-gatherer habitations include lists of activities with their concomitant artifact assemblage (e.g., Brooks and Yellen 1987; Chang 1988; Fisher 1986; Newell 1987). The analyses and discussions then account for disposal behavior, "primary" and "secondary" refuse, and the occupation of and (often) subsequent return to the site. These studies apply the basic principles of a standard approach to the organization and use of space at sites created by hunter-gatherers (see O'Connell 1987 for a broader comparative approach). These principles include a close and predictable relationship between assemblages and activity areas, the indication of activity areas by increased artifact densities or the occurrence of certain types of artifacts, and tools and site furniture that are cached at the site for reuse at a later time.

In contrast, ethnoarchaeologists' site structural analyses of agriculturalists' residences often include a list of artifacts, a general discussion of the activities occurring on these sites, and specific references to site maintenance. Indeed, frequently site maintenance is the major emphasis of these site structural analyses. The ethnoarchaeologist's task is to account for the distribution of artifacts at the site in terms of maintenance activities (e.g., Arnold 1988; Deal 1985; Dodd 1984; Hayden and Cannon 1983, 1984; Killion 1987; Kramer 1982; see also Murray 1980). The basic principles ethnoarchaeologists derive from such studies include the recognition that artifact-poor areas generally indicate the locations of activity areas, and that artifact-rich areas are usually refuse concentrations.

Disposal areas are not necessarily associated with the activity area where objects were made, food was consumed, tools were used, or refuse was generated. Activity areas may have distinct artifact assemblages associated during use, but these associations may not be evident archaeologically. Debris left in dumping zones or unmaintained space lies in a form of passive storage (Deal 1985; Hayden and Cannon 1983; Killion 1987), available for impromptu use.

During this recycling process, some artifact types become less identifiable in terms of an item's original form and use. This recycling is present among both hunter-gatherers and agriculturalists. We often consider hunter-gatherer lithic technology in terms of curation (e.g., Bamforth 1986; Binford 1979; Kelly 1988). Nelson defines curation as "a strategy of caring for tools and toolkits that can involve advanced manufacture, transport, reshaping, and caching..." (Nelson 1991). We can characterize aspects of agriculturalists' organization of technology in the same way (e.g., Nelson and Lippmeier 1993; Tomka 1993). With increased use of a residence, a household may rely more on resources already at that place in the form of refuse. Discarded objects are recalled into use as impromptu tools, thus removing them yet one more step from their original place and function (for example, Killion 1987; Tomka 1993).

Site furniture can also be thought of as curated. When the settlement system includes regular periods of abandonment and reoccupation, site furniture includes facilities that are cached in anticipation of return to the place (Graham 1993; Nelson and Lippmeier 1993). As we have seen with the Guarijío and Tuxtleco, among sedentary groups, these same facilities are permanent features, often in activity-exclusive work areas.

At times in this discussion I have tried to present an exaggerated dichotomy between the cultural systems of, broadly, hunter-gatherers and agriculturalists. Yet I think the present study demonstrates just how artificial this format can be. Mobile agriculturalists share organizational aspects with both agriculturalists and hunter-gatherers. These organizational similarities reflect commonalities in the way space is organized and used under conditions dictated by aspects of the subsistence-settlement system. The important characteristics of mobile agriculturalists' residential site structure include the placement of artifacts and activity areas, and maintenance behaviors. Artifact assemblages are removed from the activity area during site maintenance. Boundaries of activity areas are often indistinct because there are few specialized or formalized activity areas. Generalized activity areas predominate and site maintenance activities are well established, although not rigorous or formalized. Residences are stocked with the basic equipment and facilities necessary to perform most activities anticipated during their occupation; the need for personal gear and valuable tools is also presumed, and these items are included in the assemblage a household brings with them to the residence.

The Site Structure of an Agriculturalist's Residence

Some important factors influencing the organization and use of residential space are climate, topography, subsistence practices, and intensity of occupation. Climate dictates the need for sheltered lifespace. Topography can influence not only the arrangement of residences and workspace within that residence (as the terracing of some Guarijío residences attest), but also the dispersal of refuse around it. The nature of subsistence procurement (most broadly, food production or food collection), and the processing requirements of those foods for storage and daily consumption are major elements impacting the layout of residences. Intensity of occupation refers to occupation length and number of inhabitants of a place. An individual's spatial requirements are different from those of a group. In the latter case, synchronous access to the same facility or activity space by more than one person requires modifying the spatial extent of a given individual's activity, scheduling the use of activity space, and accommodating nonparticipants (for example, children or sick household members) within that space. Understanding the relationships between these factors and their material consequences will enable us to better interpret the archaeological record and the prehistoric systems that produced it.

Sheltered Lifespace

The requirement of sheltered lifespace, and how it varies with climate, subsistence practices, and length of occupation is a critical factor in the issue of space use. The majority of tasks performed at residences of the three agriculturalist groups discussed above are not shelter-specific. Nevertheless, the need for some sheltered lifespace exists in each case.

The Rarámuri require sheltered activity areas during winter months as protection from the cold for both themselves and their animals. They also need protected work space during the rainy season when heavy rains make outdoor cooking and sleeping impractical. The Guarijío also require sheltered lifespace. While winters are milder in this area than in the Sierra Tarahumara, they are sufficiently cool to necessitate the additional warmth of a hearth or brazier, or shelter within a structure. During warm weather, temperatures can be uncomfortably high. Roofed-over activity areas are particularly important at this time of year; in carrying out a number of tasks, household members choose specifically between areas of sun or shade. The Guarijío perform any food processing and preparation activities requiring specific facilities within the shaded kitchen, where the facilities are located. The Tuxtleco need for sheltered lifespace is somewhat different. Rather than the Rarámuri requirement of protection from the cold, or the Guarijío need for shaded space, the Tuxtleco endure a heavy tropical wet season, and require protection from the rain for both themselves and their goods.

Work Space

As occupation at a site intensifies (that is, as length of occupation or number of occupants increases, or as the available area for use decreases), space becomes more specialized; activity areas become more formalized and maintenance activities increase. Studies of hunter-gatherers in the process of sedentarization reflect this trend, as do studies of variability in site function, length of occupation, and reoccupation within a given group (Brooks and Yellen 1987; Hitchcock 1982, 1987; Kelley 1986; Yellen 1977). The study presented in this book demonstrates that we need not look at the gross dichotomy of hunter-gatherers v. agriculturalists to find important and meaningful variations in site structure.

The more often an activity occurs, the more likely it is that it will be performed in the same place. Following this localization of activity (or class of activities), if intensity of occupation continues to increase, the next phase in specialization of activity areas is the exclusion of other activities from that space (Anderson 1982; Arnold 1988).

The focal point of the residence for the Rarámuri, the Guarijío, and the Tuxtleco is the

Structural Activity Area. This space comprises the major structure(s) at the residence. Minimally, it shelters sleeping, storage, and food processing activity areas. Tasks are not only performed at indoor work areas, however, and the Structural Activity Area also includes the area immediately surrounding the structure(s). Both the Guarijío and Tuxtleco segregate some kinds of activities within this area, dividing the space into specialized work places.

The general pattern of activities centering around a Structural Activity Area is not restricted to agriculturalists. Although it is most well-defined among groups whose subsistence-settlement organization requires long-term occupation of the residence, this pattern also occurs where sheltered lifespace is used extensively.

For example, the distribution of activity and refuse areas at winter residences for ethnographic groups living above 55 degrees latitude follows many of the same patterns noted for sedentary agriculturalists (Graham, Raish, Sebastian 1982). During the severe winters above this latitude, outside activity areas are infrequent and, when they occur, placed quite near to the dwelling. Dumps are also close to the dwelling. Maintenance inside the Nunamiut winter house involves both sweeping and stockpiling trash, which is then dumped in the refuse areas just outside the structure's entrance (Binford, personal communication, 1982). At residences occupied during other seasons of the year, activity areas are much more extensive, and refuse tends to be much more diffuse, or dumps are removed from the major activity loci.

Among most cases of non-sedentary groups, regardless of their subsistence economies, a simpler, less specialized pattern prevails. Among groups occupying residences for relatively shorter periods of time, or where the inhabitants of the site include a number of nuclear families or other groupings, the organization of the basic components of sheltered lifespace and activity areas can be quite different. Examples of such differences include those described by Yellen in his discussion of the "Ring Model" for the !Kung (1977:125-130) and by Fisher and Strickland in their analysis of Efe Pygmy campsite components (1989:Figure 1, p. 475).

When people differentiate between kinds of space, and creating specialized activity areas, one of the first divisions separates out the placement of cooking and food processing from other activities; food processing is one of the most common activities occurring on residential sites. The more frequently an activity occurs, the more often it will occur in the same place. Because many food processing tasks require facilities, this segregation of food processing activities into specialized work areas apart from from other activities is not surprising.

The Rarámuri have areas used specifically for food processing, for example the walled activity area in Residence 18 (Figure 23). The difference between this activity area and the Guarijío kitchen or the Tuxtleco Cooking Zone is that the *exclusiveness* seen in the latter cases does not exist in the Rarámuri case. That is, while there are exceptions when Guarijío and Tuxtleco kitchens are used for sleeping areas, those areas used for food preparation among the Rarámuri also are used consistently for many other activities. At different times during the course of my fieldwork, the walled activity area at Residence 18 also housed a loom, a sleeping platform, sprouting corn used in making corn beer, and the jars in which the beer was fermented and served.

Increased length of occupation at a residence results in more specialization of activity areas. This specialization is a consequence of redundancy of activities at that place and the solution to the problem created when residents must schedule a variety of activities within the same work space. But if this is the case, why doesn't the site structure of the most permanently occupied residence in Rejogochi resemble that of less mobile agriculturalists like the Guarijío?

Residence 16 (Figures 21 and 32) is one of the few examples of a permanently occupied residence in Rejogochi. Prior to 1977, the residence was occupied by Cruz's maternal grandmother. After her death, Cruz, Irene, and their son moved from Cruz's natal residence, which was approximately 75 m upstream, to Residence 16 (Hard and Merrill 1983). The history of Residence 16 prior to 1977 is not known, but Cruz and Irene added Structure B to the dwelling when they moved there. Every year, the household remained in residence throughout the winter. The household's sheep and goats shared a barn with those belonging to Cruz's natal household, water and firewood were sufficiently close, and the dwelling itself was comfortable during inclement weather. The family did not move during the agricultural cycle, either, as Cruz and Irene cultivated land only in Rejogochi. Thus, from 1977 until Cruz's unexpected death in late 1983 or early 1984, the household lived year-round at Residence 16.

Residence 16 does not appear to be substantially different from residences in Rejogochi occupied seasonally as the main residence, or only during the agricultural cycle. The arrangement of facilities and activity areas at Residence 16 does not vary significantly from that found on other Rarámuri residences, either in interior or exterior space. I made timed behavioral observations during its occupation in late winter/early spring 1983. These, combined with my subsequent experience and observations in 1987/88, indicate the use of space at this residence was not noticeably different from that observed at other residences. The only apparent difference between this and other habitations is that the amount of trash here was considerably greater than at residences occupied for shorter time periods, or periodically over the course of several years.

In Rejogochi, the use life of dwellings, or their structural components, often extend beyond the occupation span of a residence. People also move between residences to the extent that the notion of a "permanent" occupation of a Rarámuri residence is not a productive one. Clearly, length of occupation and relative permanence of occupation are not the only factors influencing the arrangement of residences, at least among some non-mobile agriculturalists. *Anticipated* length of occupation plays an important role in influencing how residential space is organized initially and subsequently used (Kent 1991; Kent and Vierich 1989; McGuire and Schiffer 1983; Wilk and Rathje 1984).

I think several considerations played a role in how Cruz and Irene organized and used their residential space, and why it does not differ substantially from that at other homes in Rejogochi. Cruz's grandmother established a pattern of space use while she lived at the residence. Certainly, it is easy enough to reorganize a Rarámuri residence (one of the major points of this study), and the addition of Structure B on the site would have disrupted the pattern, anyway. But this established arrangement represented the starting point for future modifications Cruz and Irene would choose to make. Household 8 consisted of a young married couple and their small child. While Cruz and Irene had had several children, all but one of them died when they were quite young. The intensity of occupation at this residence, although greater in *length* than other residences in Rejogochi, was much less in *number of occupants* than many others in the community. Scheduling constraints a larger household would have faced never arose for Household 8. Anticipated residential mobility as the standard settlement pattern, combined with an established pattern of space use and no scheduling conflicts within that pattern, probably contributed to the lack of a distinctive site structure at the most permanently occupied residence in Rejogochi.

Site Furniture (Facilities and Artifacts)

Site furniture is any item left at a site, "belonging" to that site in the sense that it is left behind when the site is abandoned in anticipation of being used when the site is reoccupied (for example, Binford 1978). The concept of site furniture was originally presented in the context of mobile hunter-gatherers who move across a landscape and return to reuse a region or a previously abandoned site. Elements of site furniture are discussed for numerous groups of hunter-gatherers and pastoralist groups (Binford 1978, 1979, 1980, 1982, 1983a; Brooks and Yellen 1987; Chang 1988; Janes 1983; and others). Obviously, site furniture is not limited to those items left at residences but rather may occur at any place in the settlement system that is returned to periodically, or where return is anticipated (Brooks and Yellen 1987; Graham and Roberts 1986). If archaeologists can identify site furniture at sites, we are more likely to make accurate inferences about settlement organization and subsistence-settlement systems in the past.

Site furniture at residential sites where return is anticipated includes a basic assemblage of tools and facilities. Random House College Dictionary (1980) defines a facility as "something designed, built or installed to serve a specific function...something that permits the easier performance of an action." Examples of facilities are grinding and mealing implements (here, metates, manos, and molinos), sleeping boards and platforms, hearths, storage structures, and other buildings. Tools and other items used in activities at the site are also part of the assemblage of site furniture, such as cooking and serving vessels, storage containers, utensils, and subsistence tools, for example, digging sticks, plows, and hunting implements.

Among residentially mobile agriculturalists, the common rather than the uncommon items remain as on-site assemblages at the residence. Musical instruments and valuable items, if they are not taken along as personal gear, may remain at the most secure residence. For the Rarámuri, this is usually the main residence. Elsewhere, I have described the Rarámuri settlement system as one of punctuated abandonment and reoccupation, wherein the residence is left, but a return to it is anticipated (Graham 1993). Knowing the general structure of the activities to be performed during each reoccupation, the household leaves behind an assemblage of site furniture that provides it with the tools needed to accomplish these tasks. From an archaeological perspective, it is easy to imagine the assemblage of a Rarámuri residence that had been permanently abandoned: the structures have been dismantled and removed, and the remaining material is predominantly trash that accumulated during years of occupation. Perhaps a few items of value or utility that had been overlooked would also be present. But if we imagine encountering a site in a phase of punctuated abandonment in the archaeological record, the assemblage of site furniture would have remained intact (ignoring, for the sake of argument, the likely scavenging that would occur in such a context) (Graham 1993). In this case, we would expect the assemblage to be found in formal storage locations rather than in passive storage or active use areas.

Table 4 and Figure 26 (see also Figure 13), and Table 5 (see also Figure 28) list many of the items Household 2 left at Residences 5 and 4 during episodes of punctuated abandonment. Household 2 treated the two residences slightly differently in its abandonment of these residences. In the case of the rockshelter (Residence 4), items left in anticipation of reoccupation remain at the residence, cached out of sight or in protected areas normally used for other activities (for example, inside the dwelling, but not necessarily along the margins of the structure, the usual storage location during occupation). Residence 5 is the household's main residence. Though abandoned for the season, it is returned to periodically for additional food supplies, specialized tools and utensils, craft items, and to carry out specific activities. During the household's absence there is little or no curatorial maintenance. The interior activity space of Structure A (a protected structure with a locking door) becomes a storage area, but in Structure B the facilities remain unobstructed and the activity areas are intact (Figure 26).

The location of items in the residence is another possible means of identifying the relative use of a residence—whether or not the site is part of a system that is organized to be residentially mobile. Such a system is identifiable only when a number of residences having some variability in the relative locations of parts of the assemblage are located. Thus, I know that "formal storage" of Rarámuri ceramics involves turning pots upside-down and placing them in unused areas, as is the case with the jars in front of the stove in Structure A during the abandonment phase of Residence 5 (Figure 26). Pots found like this imply the use of the dwelling for storage rather than for occupation. The extent to which these kinds of distinctions in space use and material correlates might translate to the archaeological record requires further exploration.

As noted in previous chapters, another important difference between residences of mobile and non-mobile agriculturalists is the relative permanence of facilities. The development and use of permanent facilities reflects increasing investment in a residence that correlates with permanence of occupation. We also expect an increase in specialization of activity areas with greater relative permanence of occupation. Rarámuri facilities are moved more easily and frequently than are those of the Tuxtleco or the Guarijío. While facilities used for food processing and preparation are found in the kitchens of Tuxtleco and Guarijío residences, there is no investment in a single cooking facility or a single cooking area among the Rarámuri. Among less mobile agriculturalists, relatively permanent residential space is more structured and activities are separated into discrete loci. Facilities are built into their specialized work place. These include cooking facilities (built hearths and ovens) and mealing facilities (the metates and molinos set on tables or posts).

Arnold's (1988) discussion of permanent facilities is relevant here. In his study of the potters of Los Tuxtlas, Arnold proposes a model of ceramic production where the use of permanent firing facilities (kilns) is selected under conditions of spatial constraints. In presenting a general model of space use Arnold suggests that, as a result of the redundant performance of an activity, it is located in the same place, eventually to the exclusion of other activities (Arnold 1988). Thus, generalized activity space first becomes a specialized activity area, and then an activity-exclusive area. Arnold then says:

> Once this specific work area has been established, further intensification would select for technological change rather than behavioral reorganization.... Technological choices should thus begin to include more permanent facilities, which reduces the options of spatial flexibility. Activities thus become entrenched within certain areas of the house lot. (Arnold 1988:251-252)

That is, permanence of facilities is a product of greater investment in a place or activity, perhaps as a way of maintaining the specialization of the work area. Subsequently, increased specialization takes the form of technological reorganization (for example, kilns and cooking facilities), and results in even greater permanence.

In this context, it is interesting to look at examples of the nature of facilities, their placement, and relative permanence of facilities in other subsistence-settlement systems. The Shirdasht pastoralists of western Iran (Watson 1979) seemingly challenge the suggested relationship between continued occupation, specialization of activity areas, and increased investment in permanent facilities. The members of the community of Shirdasht are Kurdish speakers. Traditionally, they were sheep and goat herders. Beginning in the 1950s, they were coerced by the village landlord into growing wheat and barley. Each year in May, the entire community leaves its mud and stone dwellings and travels with its tents and herds into the mountains. Arriving at the first of their summer camps, Duzaray, they set up their tents (Figure 45). They live at Duzaray for about one month before moving again, higher up the mountain. Residences are established in the same place year after year (see Watson 1979:Plate 9.5, p. 251):

> Before the black-tent homes are set up at Duzaray, the tent sites can be easily picked out because of the stone-lined hearth and the (sometimes rather tumbled) stone platform [for sleeping and storage] that mark each one. (Watson 1979:263)

The Shirdasht occupy the residences at Duzaray for only a few weeks out of the year. When they leave, not only do they abandon the residence, but they dismantle the dwelling (tent), as well. As a more extreme example of residential mobility than any offered by the Rarámuri, how do we account for the relative permanence of the cooking and sleeping facilities at these sites?

I think that the Shirdasht pastoralists, though quite mobile in their transhumance, invest in building their permanent facilities because their residential mobility essentially is invariable. Each year they return to the same camps and reconstruct

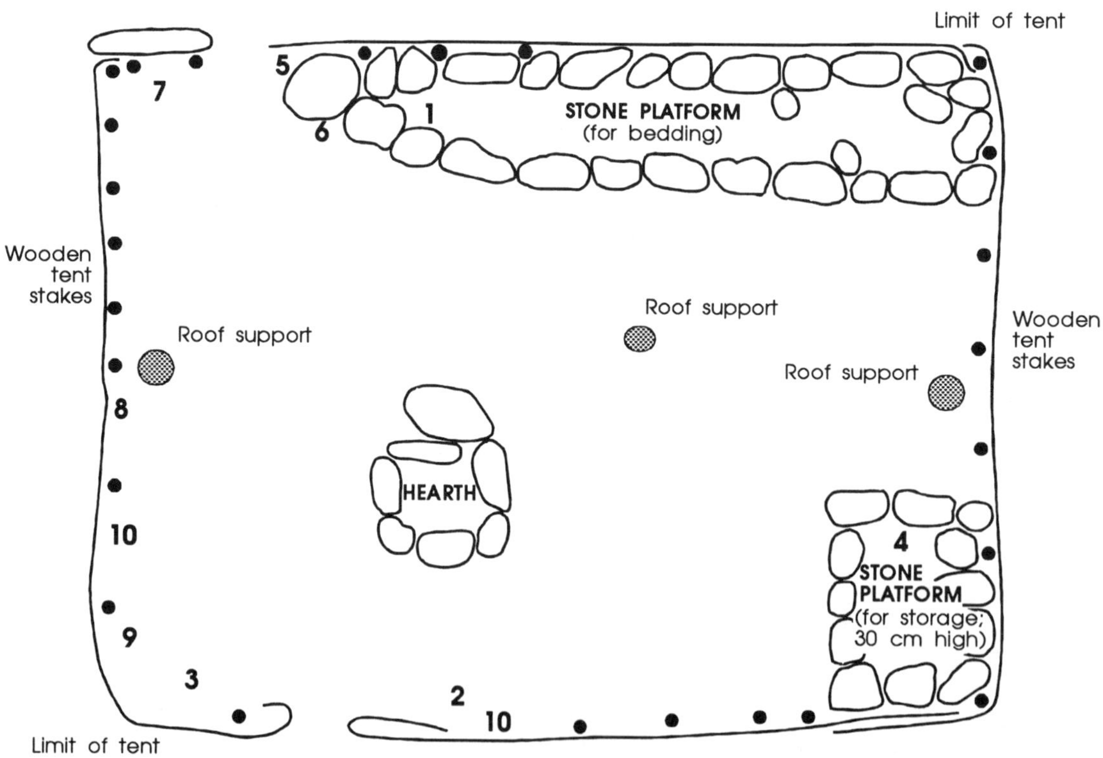

Figure 45. Interior of Shirdasht summer tent (after Watson 1979:266, Figure 10.3). 1 = rifle, 2 = pot, 3 = twig broom, 4 = basketry tray of greens, skin bag of water, two tin bowls, samovar, pot of *dugh* (buttermilk), skin bag of *rün* (clarified butter), 5 = bread dough pan (covered with cloth and basketry tray) sitting on a flour sack, 6 = brass bowl containing tea pot, glasses, and saucers, 7 = gunny sacks and saddle bags, 8 = binoculars hanging from a tent stake, 9 = metal flour sieve hanging from screen, 10 = rag draped over screen.

their dwelling in the same place they have every year before. Thus, although highly mobile (each family makes approximately five residential moves per year), the Shirdasht nevertheless repeatedly reoccupy the same houses on their transhumant route. This settlement organization can be contrasted with the Rarámuri pattern described for Rejogochi. In the latter case, occupation of a residence may vary over the years, and the role of a given residence within a household's mobility system also may change. The Rarámuri settlement organization argues for less investment in permanent facilities because of the more flexible settlement arrangement in which site function, duration of occupation, and periodicity of occupation may vary considerably.

The Shirdasht may have a more intensive use of dwelling space than do the Rarámuri, too. If the activities performed within the confines of the tent, or the number of occupants, cause activity scheduling conflicts, then I would expect the Shirdasht to adopt the use of activity-specific or activity-exclusive areas (as discussed above for winter residences above 55 degrees latitude). These would be indicated by the built-up facilities that remain in evidence whether or not the residence is occupied.

Increased labor investment is related to length and expected duration of occupation, and stability of site function. With these, we can expect to see investment in facilities. The Rarámuri and Shirdasht pastoralists share a similar mobility strategy to the extent that households repeatedly return to their residences. As the above discussion and previous chapters indicate, there are definite archaeological consequences of these systems requiring further investigation. The three cases discussed at length in this study include domestic animals in their subsistence-settlement organizations. However, the incorporation of animals into these modern-day examples should not prevent archaeologists from exploring the conditions under which similar mobility patterns may have occurred prehistorically.

A study of the land use practices of Navajo families using the south McKinley Mine lease area in the southwestern United Stated is another provocative example of pastoralist-farmers whose variant of the settlement organization includes families reoccupying a number of residences during the course of the year and over a number of years (Kelley 1986). Kelley presents a detailed ethnohistorical analysis using interview data, archaeological site records, informant data about archaeological sites, and historic documents. Using these data, she traced households and their land use areas throughout the McKinley Mine lease area from before 1881 to the 1980s. She recorded individual household histories, identified and mapped specific sites mentioned by informants, and interviewed the inhabitants of these residences. The result is an in-depth discussion of Navajo land use during the last century as economic and mobility opportunities and conditions in the lease area have changed. Within the McKinley Mine lease area, space has become increasingly constrained as population has grown, and herding, ranching, farming, and mining activities intensified. I am limiting my discussion of Kelley's findings to her data from the Railroad era (1881-1930) through the Grazing-Regulation era (1930-1950).

Kelley recorded a change in the Navajo residence pattern over time; households occupied residences for longer periods of time as mobility became more restricted (Kelley 1986). Places used during the annual round became more function-specific, and a shift towards craft specialization (weaving, blacksmithing, jewelry making) resulted in increased specialization within activity space at residences.

Some interesting site structural changes in residences in the lease area grew out of the modified subsistence-settlement organization. Figure 46 shows the percentage of these facilities occurring at Navajo residences (homesteads only) during the different eras. Kelley lists forked-stick hogans separately from the other Railroad-era sites because they occur almost exclusively in the early part of the Railroad era, and are consistently "more traditional" residences. Figure 46 shows a substantial increase in the occurrence of certain facilities on residential sites in the different eras. The range in the overall number of facilities at these sites also increased as a result of occupation episodes at residences becoming longer and more frequent than previously. Whereas the number of components on forked-stick hogan sites was nine, during the Railroad era, the number of facilities ranged from eight to ten, and in the Grazing-Regulation era from 12 to 34 (Kelley 1986).

In summary, the placement of facilities, their arrangement in space, and the relative investment in them (reflected in their permanence) are informative of settlement organization. As people become more settled, both as they occupy their residences longer, and as their mobility options are constrained to fewer and more function-specific places, they will invest more in their facilities, and

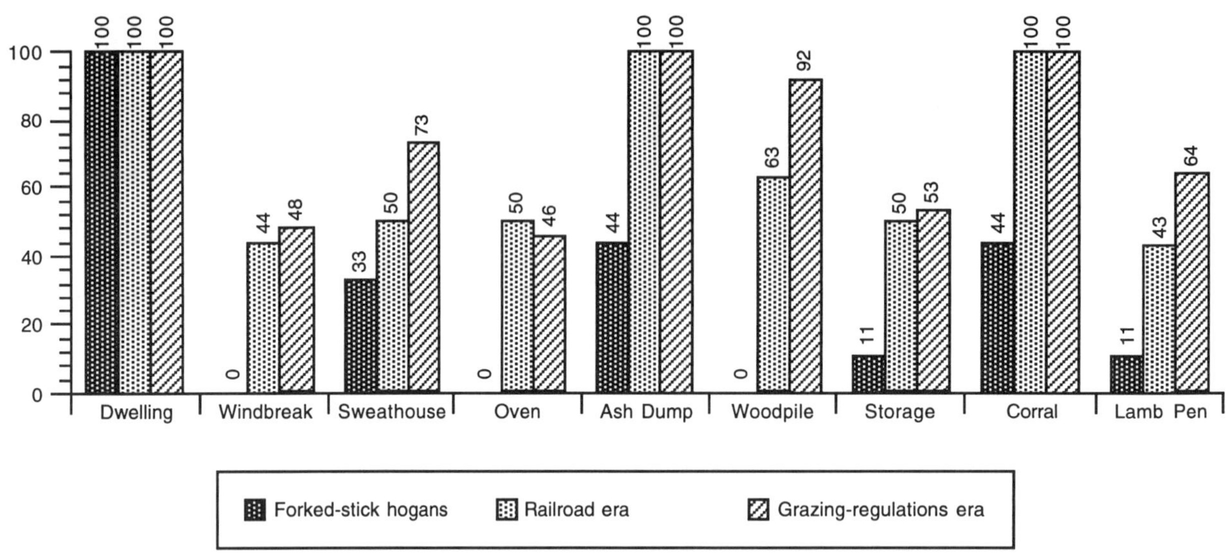

Figure 46. Percent of features at Navajo homesites (from Kelley 1986:Tables 8.3 and 13.2).

their activity areas will become more specialized. As space becomes more constrained and activity areas more specialized, scheduling pressures will be alleviated through permanent facilities, often the result of technological reorganization.

The *permanence* of facilities is thus informative about settlement organization. Moreover, because many movable facilities are put in formal storage during abandonment, their occurrence and placement at sites also reflect settlement organization. The Rarámuri often store smaller items such as the molino inside a storage structure or dwelling. They simply turn larger items like metates and cooking pots upside-down (Bennett and Zingg 1935:80), or cache them (they hide the food processing equipment in anticipation of temporary reoccupation in Basíhuare).

Refuse Areas

Without a doubt, maintenance behavior is one of the most critical classes of activities to take place at a site. This is true not only for the occupants of a residence, but for the archaeologist who encounters that residence as a site. The majority of artifacts with which archaeologists work are discarded items, no longer of practical value or use in the living system. The implications of this are twofold. First, archaeologists use these discarded fragments to make conjectures about the use of complete tools and other items in the context of the system that produced them. Secondly, and what is of concern here, we must understand something of the behavioral context in which *trash* actually is generated and distributed. Without such an understanding, we run the risk of making inferences about the past without being aware of the constraints and biases inherent in the distribution of materials. Although the discussion here concerns residential sites, much of it also applies to other kinds of sites.

A number of studies documenting maintenance behaviors attest that they depend on length of occupation, types of processing activities, and degree of specialization of activity areas, (for example, Binford 1983b; Hitchcock 1987; O'Connell 1987; Yellen 1977). While a wide range of maintenance behaviors occur at residential sites, the relative frequency of the different kinds vary as a function of the degree of maintenance practiced. From an archaeological perspective, discard behaviors generate more of the material record found on sites than does curatorial maintenance. As occupation at a residence intensifies, so do maintenance activities. A major component of most ethnoarchaeological studies of sedentary agriculturalists is the discussion of site maintenance behaviors, the resulting distribution of materials, and the possible implications in terms of an archaeological record. Hunter-gatherer residences also require maintenance, but when occupation intervals are short, or processing and other activities are not intensive, neither the investment in maintenance nor the rearrangement of refuse is as great as at sites with long-term occupations. In fact, one "maintenance" strategy among the Alyawara of Australia is to shift the occupation a short distance to a cleaner area (O'Connell 1987:88).

Discard behaviors entail removing useless, unwanted, or worn out and broken objects from the work space. Discard behaviors include many common clean-up activities. Sweeping, dumping, tossing, and placing, all remove debris from an area, relocating it elsewhere. Another class of maintenance, "curatorial maintenance," serves to keep work space free of valuable or essential equipment by the return (replacement) of tools to their proper storage location. Curatorial maintenance activities are incorporated into other aspects of space use; for example, whether an area is a specialized or generalized activity area is defined in part by the extent to which tools are put away after use, and the nature of each tool's storage location. Discard maintenance, however, has a direct association with the development of refuse areas.

The various forms of discard maintenance that a household uses relate to the need to keep activity space free, and the amount of space that can be given over to refuse, as well as to general considerations of hygiene. When residential space use is not constrained by the extent or frequency of the activities carried out there, by the physical limitations of the space, or by the number of people inhabiting that space, discard maintenance can be infrequent and generalized. Sweeping and tossing are sufficient for keeping activity areas clear and ready for other tasks. When space becomes constrained by a relative increase in the density of activities occurring in (or people inhabiting) an area, discard maintenance behaviors become more formalized, and dumping and stockpiling are common strategies. As noted before, these strategies are not peculiar to non-mobile agriculturalists. In response to increased intensity of occupation at winter residences, the Nunamiut use many of the same maintenance tactics described for the Guarijío and Tuxtleco.

Different discard behaviors have distinct material assemblages that archaeologists might be able to monitor. Sweeping is one of the most common methods of site maintenance used. Sweeping results in the removal of all large, and many small, artifacts from an area. Characteristically, swept debris is a mixture of various artifact types in a range of small sizes. A number of items, especially very small items, are not picked up by the broom, however, and remain in the activity space. According to Dodd (1984), this debris clusters in the margins of structures and around various objects where sweeping is difficult. These items become embedded in the surface of the activity area due to trampling and other activities (for example, wetting down activity space as part of the upkeep, grinding meal with water, or washing dishes).

Dumping involves consolidating trash before carrying it to a refuse area for disposal. The results of this behavior, at least initially, are discrete concentrations of debris. Trash discarded in dumping episodes frequently includes larger-sized materials than swept debris (ash is an obvious exception). Dumping behavior results in specific places (dumps) at the residence. Dumps typically occur *within* refuse areas, but under extreme conditions, they may be the primary refuse area.

Tossing is an irregular, but not necessarily an infrequent, means of discard. Tossed refuse ends up at some distance from the maintained work space—beyond the interface between the Swept Trash Area and the more dense refuse areas. Larger and heavier items seem to be tossed more frequently than others (Binford 1978). They add to the general character of the refuse area where several maintenance strategies overlap—sweeping, dumping, stockpiling, and tossing.

As discussed above, another form of discard is stockpiling, or "placing." Placed items have the potential for additional use at a later date. The term "provisional discard" (Deal 1985; Hayden and Cannon 1983) is often used to refer to what others have called "placing" (Binford 1978), and also "passive storage" (Binford 1977, 1979; Chang 1988). Provisional discard sets aside potentially recyclable refuse for the eventuality of drafting it back into use. As the numerous references suggest, this form of discard seems to be a common means of storing and recycling items. Placing is sometimes an irregular form of discard, but the Tuxtleco offer the most extreme variation of this behavior, as they actually create stockpiles of raw material stores.

In this context, stockpiling is an elaborate form of discard. Both Killion (1987) and Arnold (1988) found when residential space use is more intensive (that is, when more activities and kinds of activities are occurring at a residence), refuse areas become smaller and more concentrated. Rather than simply sweeping debris out of the way, it is collected into concentrated refuse piles. This form of maintenance allows for a relatively greater amount of activity space by significantly concentrating household refuse.

Arnold (1988) and Killion (1987) have found that surface and excavated samples of refuse areas generated under the different maintenance behaviors have distinctive assemblages. Although the patterning in the "archaeological" assemblage breaks down on Tuxtleco sites occupied for over 30 years, the implications of these findings are intriguing.

As part of an overall system of site structure, maintenance activities vary between systems. If these activities also have distinct material assemblages, a better understanding of the nature and distribution of refuse is another means to identify and interpret archaeological data on settlement organization.

Monitoring these assemblages requires careful archaeological recovery methods. The range of variation in size and weight of artifacts over an area might indicate the assemblage was generated from sweeping or dumping practices. Unless we approach the archaeological record with the question, "Can we distinguish between swept trash, dumps, and maintained space?" there is no way to know if we can infer maintenance behaviors, and their relation to settlement organization, for a prehistoric system. In light of the alterations by geologic processes that assemblages undergo, it is particularly important to address our ability to distinguish between taphonomic and organizational processes.

Changing Our Approach to the Archaeological Record

Identifying site structural properties of different kinds of subsistence-settlement organization allows us to focus on important variability in the archaeological record. As this study shows, maintaining a perspective contrasting hunter-gatherer with agriculturalists, or opposing mobility and sedentism in our interpretations of the past is of limited utility. As archaeologists, we have been aware of this limitation for some time, and have had ethnographic evidence supporting the inadequacies of dichotomizing adaptations. But we have not always modified our methods of data recovery and analysis to reflect our growing understanding of cultural systems. Nor have we successfully incorporated this understanding into robust models of prehistoric subsistence-settlement systems. Our methodology has continued to limit our understanding of the past.

In various parts of this study I have discussed the nature of activity areas and how the degree of specialization of those areas is related to competition for work space. As length of occupation increases, competition for work space also increases, because more activities are being performed at the site. Specialized activity areas are created. The purpose of these areas is to separate different classes of activities, including those tasks requiring space for long-term projects, or a place to accomplish work without interruptions. Where intensity of occupation is high, permanent features become a part of these specialized activity areas, and the areas themselves are used exclusively for a particular task or class of activities.

Important organizational changes in the assemblage of site furniture occur as groups become less mobile. If we can identify relevant facilities, then we can monitor the changes in mobility through time or across cultural systems. Food processing and preparation facilities, like grinding implements and hearths, are good examples of those we can monitor (e.g., Nelson and Lippmeier 1993; Schlanger 1991). We also need to look at permanent facilities, like the Tuxtleco kilns, Guarijío firestands, and Shirdasht platforms and hearths, to see if we can recognize permanent and movable variations of facilities. Like Arnold (1988), we can work in an ethnoarchaeological context to better understand the use of these facilities.

The ability to distinguish between different kinds of maintenance behavior through variations in trash assemblages is another valuable indicator of the nature of the occupation of a residence. We see an increase in site maintenance among mobile agriculturalists, relative to many hunter-gatherers, of a magnitude such that, as with other more sedentary people, there is little relationship between locations of an artifact and the place where it was used. Nevertheless, while maintenance behaviors vary between mobile and non-mobile groups, these differences often relate more to intensity of occupation (e.g., length of stay or number of people and amount of available work space) than to subsistence practices. If we want to map changes in types of refuse areas at a site, we need to excavate extensive contiguous areas in small units. This strategy needs to be implemented for a range of site types and time periods so we can identify and monitor the range of variation present in the archaeological record and how it reflects differences in settlement organization.

The Rarámuri as mobile agriculturalists are easily distinguished from many hunter-gatherers in their organization and use of space. It is more difficult to show differences between the Rarámuri and other, less mobile, agriculturalists. One organizational difference between mobile agricultural and more sedentary agricultural systems is the distinction between their approaches to agricultural land use. The Tuxtleco option to use field houses rather than the Rarámuri practice of occupying multiple residences is one example. Other differences include the distribution of agricultural fields, the amount of time invested in tending and harvesting crops, and the organization of agricultural labor. Land use patterns are reflected in some measure in storage requirements, crop and food preparation activities, and length of occupation and occupation composition at different sites. Field houses, for example, might have less storage space, and a smaller ceramic assemblage than other types of residences.

The notion of increased specialization of activity areas and increased formalization of maintenance concomitant with increased intensity of occupation is critical to the understanding of the organization and use of residential space. The most common factors resulting in "intensification" of an occupation are increased length of occupation and increased number of occupants at a residence (Oswald 1987; Rodman 1985).

Intensification of space use has direct material consequences. With specialization of space we can expect to see delimiters of that space. Therefore, can we infer specialization of activity areas based on the presence of multiple structures on a residence? Arnold's (1988) model of site structure proposes that commonly occurring tasks are localized to the exclusion of other activities when there is increased scheduling pressure for various activities. Required facilities then become permanently fixed (and often technologically different from earlier forms). Dwellings are, in many respects another form of facility (Hunter-Anderson 1977). It is productive to look for evidence of variations in room function when multiple structures are present at a residence. The Rarámuri quite frequently occupy two dwellings, but while they use them differently from each other on occasion, there is no consistent specialization of the space. The artifact and feature assemblages in these rooms reflect this lack of specialization for the Rarámuri. The presence of this aspect of spatial organization should indicate groups with relatively greater residential occupation intensity. The interior arrangement and contents of structures originally used as dwellings, but later given over to storage is different from that of occupied dwellings. (Structure A of Residence 5 is a seasonal example of this change in function; Structure C of Residence 18 is an example of a longer-term change.) While the nature and placement of storage facilities are important considerations when trying to understand subsistence-settlement organization, we need to be able to identify the presence or absence of other specialized areas.

How might we infer a *generalized* activity structure within a dwelling? Dodd (1987) suggests the arrangements of hearths and other fire-related facilities reflects a generalized use of space. In a cross-cultural study, he found centrally located fire facilities are multi-functional—that is, required for both cooking and heating. When these facilities are no longer used for warmth, they are placed against a wall, frequently in a corner (Dodd 1987).

Incorporating site structural analyses and a theory of lifespace in our investigations of the archaeological record, we have the ability to interpret our findings better. Perhaps more critically, site structural studies allow us to identify meaningful variability, and to begin to understand that variability in terms of cultural process and adaptation.

Once we have identified the kinds of variability we must monitor in order to answer our questions about the past, we have to modify our approach to the archaeological record itself. We must learn to frame our questions about the past in light of our general knowledge of cultural systems. Finally, we have to rethink our data recovery methods and analytical techniques so that we can better identify and interpret the organizational, as well as taphonomic, processes that result in the archaeological record.

Chapter 6
Conclusions

At the beginning of this book, I posed several questions regarding Rarámuri settlement organization and our ability to infer it from the archaeological record. These questions include: When residential moves are made, but site function remains the same at each residence throughout the year, what does it mean to speak of season of occupation? That is, how do we incorporate a notion of seasonal change without a concomitant site-functional or subsistence change? I also asked: How far apart must seasonally occupied residences be before they represent a system of biseasonal mobility? Given the relatively short distances the Rarámuri move between residences, can we really call them mobile agriculturalists? This book provides some suggestions as to how we might begin to identify means of determining that the Rarámuri are mobile throughout the year. However, the questions raised here merit further discussion because they deal not only with our ability to understand mobility as a component of agriculturalist systems, but also with how we go about identifying mobile systems archaeologically.

We have tied our indicators of mobility in archaeological contexts to other aspects of settlement organization—most often, seasonal occupation and site function. When our recognition criteria for those aspects don't indicate impermanence of occupation, and when our models don't include the likelihood of mobility as part of the settlement organization, then we will fail to recognize variability resulting from residential mobility for what it is. Chances are, we will try to interpret that variability in terms of other aspects of the cultural system instead.

The Rarámuri offer us an example of a people who don't meet the standard criteria we use commonly to infer mobility from the archaeological contexts. Their moves are not linked to seasonal resources that we usually monitor, nor do they have the expected activity differences between residences occupied at different times of the year or for different durations. Our challenge is to develop the means by which we can interpret their settlement pattern accurately.

The ethnoarchaeological study presented here demonstrates that site structural analysis of residential sites is one means by which we can better our understanding of settlement organization. The organization and use of space at a place is dependent on the activities that occur there, the space available, the length of occupation, and the number of occupations. Intensity of occupation is the term used here to refer to the combination of these factors. With increased intensity of occupation at a residence, space use becomes more formalized. For that reason, groups with a more sedentary settlement system have more activity-specific and activity-exclusive work areas at their residences, and more permanent (built-in) features and facilities, than do groups whose settlement systems incorporate relatively frequent residential moves.

Agriculturalists, hunter-gatherers, pastoralists, and groups with mixed economies share the same principles of space use. When they face analogous spatial constraints, they respond with similar arrangements of activity areas and residential space. Site structural studies focusing on the residence as a whole abound for settings where intensity of occupation is generally low—that is, among hunter-gatherers (e.g., Binford 1978, 1982, 1983b; Brooks and Yellen 1987; Fisher and Strickland 1989; O'Connell 1987; Yellen 1977). When intensity of occupation increases (here, among agriculturalists), site structural studies tend to omit detailed discussions of the residence per se, and focus on the treatment of refuse, or aspects of specialization at the site (e.g., Arnold 1991; Deal 1985; Hayden and Cannon 1983, 1984; Killion 1987; Tomka 1993). I suggest that this shift in emphasis reflects organizational differences between "low" and "high" intensity occupations, rather than differences in economy or settlement organization per se.

In order to evaluate this suggestion, however, we need additional studies on the general ways in which groups of people organize and use space. With a few exceptions (e.g., Arnold 1988; Dodd 1984, 1987, and 1992; Killion 1987), complete information on the placement of activity areas, refuse areas, and facilities necessary to compare residential space use is lacking from published ethnoarchaeological studies of agriculturalists.

We need to consider residential site structure in terms of the nature of the activities that occur there, the number of people at the site, and its length of occupation, as well as the inhabitants' general subsistence economy. When we do so, we have the opportunity to look at the archaeological record (and cultural systems in general) from a different perspective. This perspective on how people organize and use their life space relates to settlement organization in a different way than the commonly used indicators of mobility, season of occupation, and site function. Site structural analyses have the potential to be a powerful tool we can use to interpret the distribution of archaeological materials across the landscape.

Underlying the questions presented at the beginning of this book and above, is the issue of our ability to recognize mobility as a component of agriculturalist systems, and to incorporate it into our models of how people use a landscape. Are the Rarámuri mobile farmers? According to some definitions, the Rarámuri are sedentary (e.g., Kent, ed. 1989; Murdock 1962, 1966). The length of occupation at residences both during phases of punctuated occupancy and overall residential life history are viewed as sufficiently long for us to consider these residences "permanently" occupied. Moreover, given that the distance between residences usually is not great, a number of people have asked me whether the system of punctuated abandonment and occupation described for the Rarámuri of Rejogochi should be represented as residential mobility at all.

I consider the Rarámuri mobile farmers. Moreover, their movement between residences, regardless of distance, or duration of residency, is not unique; other cultural groups, both past and present, participate in similar mobility strategies (see for example, Anderson 1982; Hitchcock 1982; Kelley 1986; Tomka 1993). Clearly, the Rarámuri present us with the opportunity to refine some of our concepts and definitions regarding settlement organization. If the moves described between Rarámuri summer and winter houses, or between their main residences and those inhabited during the agricultural cycle, don't fall under the definition of "residential mobility," then what are the criteria for residential moves?

Moreover, if we establish a minimum distance that must be travelled between residences, and/or a minimum duration of habitation, we still must be able to distinguish between patterning in the archaeological record resulting from mobility strategies and other aspects of subsistence-settlement systems like population, wealth and status, and so on. Regardless of whether we identify the Rarámuri as mobile or sedentary, the archaeological record resulting from such a system contains more than a single residence per household within and across communities. Each of the multiple residences is occupied in a systematic fashion during the year; both its occupation and abandonment are anticipated and planned for. Even if we choose not to consider the Rarámuri as mobile agriculturalists, we still need to characterize their settlement organization accurately. If we decide that recognizing Rarámuri biseasonal or agricultural-cycle moves isn't critical to (or possible in) understanding the Rarámuri settlement system, as archaeologists, we still face the problem of interpreting a settlement pattern that includes more sites than expected, given a model of sedentary agriculturalists, in addition to explaining the variations in assemblages that result from the pattern of punctuated abandonment. In other words, we must still deal with the material consequences to Rarámuri settlement organization, regardless of whether we choose to call them "mobile farmers" or not.

The mobility strategies described for Rejogochi households vary from year to year, as well as from household to household. While I don't think we can identify individual mobility strategies of households archaeologically, we can consider them at the more general level of a settlement system with a component of flexible residential mobility. For the Rarámuri of Rejogochi, the flexibility of the mobility strategies works within certain constraints, including the need for sheltered living areas, labor organization, social interactions, household size, and the distribution of available land.

As mobile farmers, the Rarámuri offer us the opportunity to consider the adequacy of our methods for identifying variations in settlement organization archaeologically. The ethnoarchaeological study of residential site structure presented here demonstrates our need to develop models and recognition criteria for identifying mobile agricultural systems. Studying the organization and use of space at residences is one means by which we can monitor settlement organization as reflected in the intensity of occupation, provided we modify our data recovery and analysis techniques to capture the relevant spatial and material data in the archaeological record. Perhaps most importantly, this study of Rarámuri residential site structure has provided us the opportunity to erase temporarily the boundaries between subsistence-settlement systems and consider residences in the context of the people who live at these places and how they arrange and carry out activities there.

Bibliography

Anderson, Dana
1982 Space Use and Site Structure. *Haliksa'i: University of New Mexico Contributions to Anthropology* 1:120-141.

Arnold, Philip J., III
1987 Site Structural Approach to Modeling Ceramic Production. Paper presented at the Fifty-Second Annual Meeting of the Society for American Archaeology, Toronto, Canada.
1988 The Household Potters of Las Tuxtlas: An Ethnoarchaeological Study of Ceramic Production and Site Structure. Unpublished Ph.D. dissertation, Department of Anthropology, University of New Mexico, Albuquerque.
1991 *Domestic Ceramic Production and Spatial Organization: A Mexican Case Study in Ethnoarchaeology*. Cambridge University Press, Cambridge.

Ascher, Robert
1969 Time's Arrow and the Archaeology of a Contemporary Community. In *Settlement Archaeology*, K. C. Chang, ed., pp. 43-52. National Press Books, Palo Alto.

Bamforth, D.B.
1986 Technological Efficiency and Tool Curation. *American Antiquity* 51:38-50.

Bender, D. R.
1967 Refinement of the Concept of Household: Families, Coresidence, and Domestic Functions. *American Anthropologist* 69:493-504.

Bennett, Wendell C., and Robert M. Zingg
1935 *The Tarahumara: An Indian Tribe of Northern Mexico*. University of Chicago Press, Chicago.

Binford, Lewis R.
1977 Forty-seven Trips: A Case Study in the Character of Some Formation Processes. In *Stone Tools as Cultural Markers*, R.V.S. Wright, ed., pp. 24-36. Australian Institute of Aboriginal Studies, Canberra.
1978 Dimensional Analysis of Behavior and Site Structure: Learning from an Eskimo Hunting Stand. *American Antiquity* 43:330-361.
1979 Organization and Formation Processes: Looking at Curated Technologies. *Journal of Anthropological Research* 35:255-273.
1980 Willow-Smoke and Dog's Tails: Hunter-Gatherer Settlement Systems and Archaeological Site Formation. *American Antiquity* 45:1-17.
1981 Behavioral Archaeology and the "Pompeii Premise." *Journal of Anthropological Research* 37:195-208.
1982 The Archaeology of Place. *Journal of Anthropological Archaeology* 1:5-31.
1983a People and Their Lifespace. In *In Pursuit of the Past: Decoding the Archaeological Record*, pp. 144-192. Thames and Hudson, London.
1983b *In Pursuit of the Past: Decoding the Archaeological Record*. Thames and Hudson, London.
1987 Researching Ambiguity: Frames of Reference and Site Structure. In *Method and Theory for Activity Area Research: An Ethnoarchaeological Approach*, Susan Kent, ed., pp. 449-512. Columbia University Press, New York.

Brooks, Alison and John Yellen
1987 The Preservation of Activity Areas in the Archaeological Record: Ethnoarchaeological and Archaeological Work in Northwest Ngamiland, Botswana. In *Method and Theory for Activity Area Research: An Ethnoarchaeological Approach*, Susan Kent, ed., pp. 63-106. Columbia University Press, New York.

Bye, Robert A., Jr.
1976 The Ethnoecology of the Tarahumara of Chihuahua, Mexico. Unpublished Ph.D. dissertation, Department of Biology, Harvard University, Boston.
1979 Incipient Domestication of Mustards in Northwest Mexico. *The Kiva* 44:237-256.

1981 Quelites—Ethnoecology of Edible Greens —Past, Present, and Future. *Journal of Ethnobiology* 1:109-123.

Camilli, Eileen L.
1983 Site Occupational History and Lithic Assemblage Structure: An Example from Southeastern Utah. Unpublished Ph.D. dissertation, Department of Anthropology, University of New Mexico, Albuquerque.

Champion, Jean R.
1962 Study in Culture Persistence: The Tarahumaras of Northwestern Mexico. Unpublished Ph.D. dissertation, Department of Anthropology, Columbia University, New York.

Chang, Claudia
1988 Nuayalik Fish Camp: An Ethnoarchaeological Study in Activity-Area Formation. *American Antiquity* 53:145-157.

Deal, Michael
1985 Household Pottery Disposal in the Maya Highlands: An Ethnoarchaeological Interpretation. *Journal of Anthropological Archaeology* 4:243-291.

Dodd, Walter A.
1984 The Use of Domestic Space by Sedentary Households: Some Organizing Principles. Paper presented at the Forty-ninth Annual Meeting of the Society for American Archaeology, Portland.
1985 The Composition of Gear Taken on Guarijío Trips. Paper presented at the Fiftieth Annual Meeting of the Society for American Archaeology, Denver.
1987 Factors Conditioning the Placement of Fire-Related Facilities and Refuse. Paper presented to the Fifty-Second Annual Meeting of the Society for American Archaeology, Toronto, Canada.
1992 Organizational Aspects of Spatial Structure in Guarijío Sites. Unpublished Ph.D. dissertation, Department of Anthropology, University of Utah, Salt Lake City.

Fisher, John W., Jr.
1986 Shadows in the Forest: Ethnoarchaeology Among the Efe Pygmies. Unpublished Ph.D. dissertation, Department of Anthropology, University of California, Berkeley.

Fisher, John W., Jr., and Helen C. Strickland
1989 Ethnoarchaeology Among the Efe Pygmies, Zaire: Spatial Organization of Campsites. *American Journal of Physical Anthropology* 78:473-484.

Flannery, Kent V.
1972 The Origins of the Village as a Settlement Type in Mesoamerica and the Near East: A Comparative Study. In *Man, Settlement, and Urbanism*, P. J. Ucko and G. W. Dimbleby, eds., pp. 25-53. Duckworth, London.

Foley, Robert
1981 Off-Site Archaeology: An Alternative Approach for the Short-Sited. In *Patterns of the Past: Studies in Honor of David Clark*, I. Hodder, G. Isaac, and N. Hammond, eds., pp. 157-183. Cambridge University Press, Cambridge.

Gentry, Howard Scott
1963 The Warihio Indians of Cinerea-Chihuahua: An Ethnographic Survey. *Bureau of American Ethnology Bulletin* 186, Anthropological Papers No. 65.

Gilman, Patricia A.
1987 Architecture as Artifact: Pit Structures and Pueblos in the American Southwest. *American Antiquity* 52:538-564.

Gotthilf, Vivian
1982 Maize Storage Strategies: An Ethno-Archaeological Perspective. Unpublished Master's Thesis, Simon Fraser University, Calgary.

Graham, Martha
1993 Settlement Organization and Residential Variability among the Rarámuri. In *Abandonment of Settlements and Regions: Ethnoarchaeological and Archaeological Approaches*, Catherine M. Cameron and Steve A. Tomka, eds., pp. 25-42. Cambridge University Press, Cambridge.

Graham, Martha, and Robert J. Hard
1983 Field map of Rejogochi. Map in possession of authors.

Graham, Martha, Carol Raish, and Lynne Sebastian
1982 Site Structure and Nunamiut Ethnoarchaeology. *Haliksa'i: University of New Mexico Contributions to Anthropology* 1:110-119.

Graham, Martha, and Alexandra Roberts
1986 Residentially Constrained Mobility: A Preliminary Investigation of Variability in Settlement Organization. *Haliksa'i: University of New Mexico Contributions to Anthropology* 5:104-114.

Hard, Robert J.
1983 Tarahumara Ethnoarchaeology Project Research Design. Unpublished manuscript.

Hard, Robert J., and William L. Merrill
1983 Tarahumara Ethnoarchaeology Project —Families and Fields. Ms. in possession of the author.
1992 Mobile Agriculturalists and the Emergence of Sedentism: Perspectives from Northern Mexico. *American Anthropologist* 94:601-620.

Hayden, Brian, and Audrey Cannon
1983 Where the Garbage Goes: Refuse Disposal in the Mayan Highlands. *Journal of Anthropological Archaeology* 2:117-163.
1984 *The Structure of Material Systems: Ethnoarchaeology in the Maya Highlands*. SAA Papers No. 3. Society for American Archaeology, Washington, D.C.

Hinton, Thomas B.
1983 Southern Periphery: West. In *Southwest, Handbook of North American Indians* 10, Southwest, Alfonso Ortiz, ed., pp. 315-328. Smithsonian Institution, Washington, D.C.

Hitchcock, Robert
1982 The Ethnoarchaeology of Sedentism: Mobility Strategies and Site Structure Among Foraging and Food Producing Populations in the Eastern Kalahari Desert, Botswana. Unpublished Ph.D. dissertation, Department of Anthropology, University of New Mexico, Albuquerque.
1987 Sedentism and Site Structure: Organizational Changes in Kalahari Basarwa Residential Locations. In *Method and Theory for Activity Area Research: An Ethnoarchaeological Approach*, Susan Kent, ed., pp. 374-423. Columbia University Press, New York.

Hunter-Anderson, Rosalind L.
1977 Theoretical Approach to the Study of House Form. In *For Theory Building in Archaeology*, Lewis R. Binford, ed., pp. 287-315. Academic Press, New York.

Janes, Robert R.
1983 *Archaeological Ethnography Among MacKenzie Basin Dene, Canada*. The Arctic Institute of North America Technical Paper No. 28. Calgary.

Kelley, Klara B.
1986 *Navajo Land Use: An Ethnoarchaeological Study*. Academic Press, New York.

Kelly, Robert L.
1988 The Three Sides of a Biface. *American Antiquity* 53:717- 734.

Kennedy, John G.
1963 Tesgüino Complex: The Role of Beer in Tarahumara Culture. *American Anthropologist* 65:620-40.
1978 *Tarahumara of the Sierra Madre: Beer, Ecology, and Social Organization*. AHM Publishing Corporation, Arlington Heights.

Kennedy, John G., and Raul A. Lopez
1981 *Semana Santa in the Sierra Tarahumara: A Comparative Study in Three Communities*. Occasional Papers of the Museum of Culture History, No. 4. University of California at Los Angeles, Los Angeles.

Kent, Susan
1991 The Relationship Between Mobility Strategies and Site Structure. In *The Interpretation of Spatial Patterning Within Stone Age Archaeological Sites*, T. Douglas Price and Ellen Kroll, eds., pp. 33-59. Plenum Publishing Corporation, New York.

Kent, Susan, ed.
1987 *Method and Theory for Activity Area Research: An Ethnoarchaeological Approach*. Columbia University Press, New York.
1989 *Farmers as Hunters—The Implications of Sedentism*. Cambridge University Press, Cambridge.

Kent, Susan, and Helga Vierich
1989 The Myth of Ethnic and Economic Determinism: Anticipated Mobility and Site Spatial Organization. In *Farmers as Hunters—The Implications of Sedentism*, Susan Kent, ed. Cambridge University Press, Cambridge.

Killion, Thomas W.
1987 Agriculture and Residential Site Structure Among Campesinos in Southern Veracruz, Mexico: A Foundation for Archaeological Inference. Unpublished Ph.D. dissertation, Department of Anthropology, University of New Mexico, Albuquerque.

Kramer, Carol
1982 *Village Ethnoarchaeology: Rural Iran in Archaeological Perspective*. Academic Press, New York.

Lumholtz, Carl S.
1894 *American Cave Dwellers; the Tarahumares of the Sierra Madre.* Bulletin of the American Geographical Society, Vol. III.
1902 *Unknown Mexico*, 2 volumes. Charles Scribner's Sons, New York.

Mares Trias, Albino
1982 *Ralamuli Nu'tugala Go'ame (Comida de los Tarahumares).* Don Burgess McGuire, Chihuahua, Mexico.

McGuire, Randall H., and Michael B. Schiffer
1983 Theory of Architectural Design. *Journal of Anthropological Archaeology* 2:277-303.

Merrill, William L.
1978 Thinking and Drinking: A Rarámuri Interpretation. In *The Nature and Status of Ethnobotany*, Richard I. Ford, ed., pp. 101-17. Anthropological Papers, No. 67, Museum of Anthropology, University of Michigan, Ann Arbor.
1981 The Concept of Soul Among the Rarámuri of Chihuahua, Mexico: A Study in World View. Unpublished Ph.D. dissertation, Department of Anthropology, University of Michigan, Ann Arbor.
1983 Tarahumara Social Organization, Political Organization, and Religion. In *Southwest, Handbook of North American Indians 10*, Alfonso Ortiz, ed., pp. 290-305. Smithsonian Institution, Washington, D.C.
1988a *Rarámuri Souls: Knowledge and Social Process in Northern Mexico.* Smithsonian Institution Press, Washington, D.C.
1988b Religion, Resistance, and Cultural Transformation: The Jesuit Mission Program in Nueva Vizcaya. Paper presented at the conference "Experiencing Colonialism: Politics, Ritual, and Consciousness in the New World," National Museum of Natural History, Smithsonian Institution, Washington, D.C..
n.d. Rarámuri Easter. In *Performing the Renewal of Community: Indigenous Easter Rituals in North Mexico and Southwest U.S.*, N. R. Crumrine and R. B. Spicer, eds. University Press of America, Lanham.

Murdock, George Peter
1969 Correlations of Exploitative and Settlement Patterns. In *Contributions to Anthropology: Ecological Essays.* National Museums of Canada Bulletin No. 230, Anthropological Series No. 86, David Damas, ed., pp. 129-162.

Murdock, George Peter, and Suzanne F. Wilson
1972 Settlement Patterns and Community Organization: Crosscultural Codes 3. *Ethnology* 11:254-279.

Murdock, George Peter, et al.
1962-1966 *Ethnographic Atlas. Ethnography.*

Murray, Pricilla
1980 Discard Location: The Ethnographic Data. *American Antiquity* 45:490-502.

Nelson, Margaret C.
1991 The Study of Technological Organization. In *Archaeological Method and Theory*, vol. 3, M. B. Schiffer, ed., pp. 57-100. University of Arizona Press, Tucson.

Nelson, Margaret C. and Heidi Lippmeier
1993 Grinding-Tool Design as Conditioned by Land-Use Pattern. *American Antiquity* 58:286-305.

Netting, Robert McC., Richard R. Wilk, and Eric J. Arnould, eds.
1984a *Households: Comparative and Historical Studies of the Domestic Group.* University of California Press, Berkeley.
1984b Introduction. In *Households: Comparative and Historical Studies of the Domestic Group*, Robert McC. Netting, Richard R. Wilk, and Eric J. Arnould, eds., pp. xiii-xxxviii. University of California Press, Berkeley.

Newell, Raymond R.
1987 Reconstruction of the Partitioning and Utilization of Outside Space in a Later Prehistoric/Early Historic Inupiat Village. In *Method and Theory for Activity Area Research: An Ethnoarchaeological Approach*, Susan Kent, ed., pp. 107-175. Columbia University Press, New York.

O'Connell, James F.
1987 Alyawara Site Structure and Its Archaeological Implications. *American Antiquity* 52:74-108.

Oswald, Dana
1987 The Organization of Space in Residential Buildings: A Cross-Cultural Perspective. In *Method and Theory for Activity Area Research: An Ethnoarchaeological Approach*, Susan Kent, ed., pp. 295-344. Columbia University Press, New York.

Panero, Julius, and Martin Zelnik
1979 *Human Dimension and Interior Space: A Source Book of Design Reference Stan-*

dards. The Architectural Press, Ltd., London.

Pennington, Campbell W.

1963 *The Tarahumar of Mexico: Their Environment and Material Culture*. University of Utah Press, Salt Lake City.

1983 Tarahumara. In *Southwest, Handbook of North American Indians* 10, Alfonso Ortiz, ed., pp. 276-287. Smithsonian Institution, Washington, D.C.

Powell, Shirley

1983 *Mobility and Adaptation: The Anasazi of Black Mesa, Arizona*. Southern Illinois University Press, Carbondale.

Rafferty, Janet E.

1985 The Archaeological Record on Sedentariness: Recognition, Development, and Implications. In *Advances in Archaeological Method and Theory*, vol. 8, Michael B. Schiffer, ed., pp, 113-156. Academic Press, New York.

Random House College Dictionary

1980 *Random House College Dictionary*. Random House, Inc., New York.

Rodman, Margaret

1985 Moving Houses: Residential Mobility and the Mobility of Residences in Longana, Vanuatu. *American Anthropologist* 87:56-72.

Rowed, Jay

1947 Summer Institute of Linguistics notes on file at the Arizona State Museum.

Santley, Robert R.

1985 Final Field Report, Matacapan Project: 1984 Season. Submitted to the Instituto Nacional de Antropologia e Historia, Mexico, and the National Science Foundation, Washington, D.C. On file, Department of Anthropology, University of New Mexico, Albuquerque.

Schiffer, Michael B.

1972 Archaeological Context and Systemic Context. *American Antiquity* 37:156-165.

1987 *Formation Processes of the Archaeological Record*. University of New Mexico Press, Albuquerque.

Schlanger, Sarah H.

1991 On Manos, Metates, and the History of Site Occupation. American Antiquity 53:460-473.

Sheridan, Thomas E., and Thomas H. Naylor, eds.

1979 *Rarámuri: A Tarahumara Colonial Chronicle, 1607-1791*. Northland Press, Flagstaff.

Spicer, Edward H.

1962 *Cycles of Conquest: The Impact of Spain, Mexico, and the United States on the Indians of the Southwest, 1533-1960*. University of Arizona Press, Tucson.

Tomka, Steve A.

1993 Site Abandonment Behavior among Transhumant Agro-Pastoralists: The Effects of Delayed Curation on Assemblage Composition. In *Abandonment of Settlements and Regions: Ethnoarchaeological and Archaeological Approaches*, Catherine M. Cameron and Steve A. Tomka, eds., pp. 11-24. Cambridge University Press, Cambridge.

Urban, Maria Teresa Cerqueira

1975 The Food and Nutrient Intake of the Tarahumara Indians of Mexico. Unpublished Master's Thesis, Department of Home Economics, University of Iowa, Iowa City.

Velasco Rivero, Pedro de

1987 *Danzar o Morir: Religión y Resistencia a la Dominación en la Cultura Tarahumar*. 2a. edición. Mexico City, Centro de Reflexión Teológica.

Watson, Patty Jo

1979 *Archaeological Ethnography in Western Iran*. Viking Fund Publications in Anthropology, No. 57.

Wilk, Richard R.

1983 Little House in the Jungle: The Causes of Variation in House Size among Modern Kekchi Maya. *Journal of Anthropological Archaeology* 2:99-116.

1984 Households in Process: Agricultural Change and Domestic Transformation among the Kekchi Maya of Belize. In *Households: Comparative and Historical Studies of the Domestic Group*, Robert McC. Netting, Richard R. Wilk, and Eric J. Arnould, eds., pp. 217-244. University of California Press, Berkeley.

Yanagisako, Sylvia Junko

1979 Family and Household: The Analysis of Domestic Groups. *Annual Review of Anthropology* 8:161-205.

Yellen, John

1977 *Archaeological Approaches to the Present: Models for Reconstructing the Past*. Academic Press, New York.